T0317462

Sar Image Analysis – A Computational Statistics Approach

IEEE Press
445 Hoes Lane
Piscataway, NJ 08854

Sar Image Analysis – A Computational Statistics Approach

With R Code, Data, and Applications

Alejandro C. Frery
Victoria University of Wellington

Jie Wu
Shaanxi Normal University

Luis Gomez
Universidad de Las Palmas de
Gran Canaria

Published by John Wiley & Sons, Inc., Hoboken, New Jersey.
Published simultaneously in Canada.

Library of Congress Cataloging-in-Publication Data:

Names: Frery, Alejandro C., author. | Wu, Jie, 1985- author. | Gomez, Luis (Gomez Deniz), author.
Title: SAR image analysis — a computational statistics approach : with R code, data, and applications / Alejandro C. Frery, Jie Wu, Luis Gomez.
Description: Hoboken, NJ : Wiley-IEEE Press, 2022. | Includes bibliographical references and index.
Identifiers: LCCN 2022000525 (print) | LCCN 2022000526 (ebook) | ISBN 9781119795292 (cloth) | ISBN 9781119795322 (adobe pdf) | ISBN 9781119795469 (epub)
Subjects: LCSH: Synthetic aperture radar. | R (Computer program language)
Classification: LCC TK6592.S95 F74 2022 (print) | LCC TK6592.S95 (ebook) | DDC 621.36/78–dc23/eng/20220215
LC record available at https://lccn.loc.gov/2022000525
LC ebook record available at https://lccn.loc.gov/2022000526

Cover Design: Wiley
Cover Image: © Yarr65/Alamy Stock Photo

Set in 9.5/12.5pt STIXTwoText by Straive, Chennai, India

To my husband and my friends, they are the salt of my life

Alejandro C. Frery

To my family, my friends, and my cute girl, they are treasures of me

Jie Wu

To Stella, my wife and support, and to my treasures and all, Prince Gaby and Princess Martina

Luis Gomez

Contents

Foreword by Luis Alvarez

Remote sensing is a well consolidated research area which demands efficient image processing methods. The many particularities of the Synthetic Aperture Radar (SAR) devices require highly specialized techniques for processing the SAR images. I have been invited by the authors of *SAR Image Analysis — A Computational Statistics Approach* to write this Foreword. This is a great honor that, with enormous pleasure, I accept. I know the authors well because I have collaborated with them on several occasions, precisely, in image processing techniques applied to Remote Sensing. SAR acquisition devices are remote sensing systems of great technical complexity that provide high-resolution images regardless of atmospheric conditions (rain, clouds, dust, etc.), and independently of day or night.

This book focuses on providing deep knowledge and interpretation of the SAR data from a plain and rigorous statistical point of view. The comprehensive approach to the topic proposed in this book will help readers to understand better all the relevant matters related to SAR images.

Of course, there are other excellent books for SAR imagery, however I found that this is not just another SAR book. From the first pages, one is aware that this volume has been written by true experts in the area who made an enormous effort to condense in these pages, years of study and research. As a result, the main SAR concepts needed for image processing are well discussed and nicely introduced. Every single line, figure, table, and programming line have been done with extreme care. From that, the whole book is easy to read and follow because it has been written in an academic spirit so that it can be understood not only by experts but by young students who are starting to study this subject.

As mentioned above, SAR is not a strange area for me. However, as a user of SAR imagery, I have not been particularly involved in the issues related to the acquisition systems or even in knowing the deployed SAR systems. Nor also on how to access SAR data from the different websites that offer free SAR data. I have also noticed that this occurs in other fields. For instance, deep knowledge of medical instruments is not required to be able to implement powerful segmentation algorithms for processing medical images. However, it is probable that better methods could be designed with more knowledge of the acquisition systems. In that sense, I found Chapter 1, "Data Acquisition" extremely useful because it provides such demanding knowledge but avoids the very complex mathematical tools that surround the radar signal. Readers can easily understand the concept of synthetic aperture and gain a physical knowledge of the backscatter and the main distortions associated with the SAR acquisition.

One fundamental aspect that centers most of the research in SAR images is the speckle noise or speckle. That granular pattern is quite difficult to manage and reduce. Speckle corrupts the SAR image (all SAR images!). It makes image interpretation or any post-processing operation more difficult. Correctly modeling speckle is a crucial issue for anyone intended to fully understand SAR data and to use or design competitive image processing methods.

A sound statistical approach is needed to deal with speckle, and this approach is the one found in this book. The reader will find in Chapter 3, "Intensity SAR data and the Multiplicative Model," updated and more reliable statistical distributions to deal with SAR data. All the distributions are nicely described, justified in the context of SAR (for instance the G0 distribution to model textured and extremely textured areas), and with a clear perspective of applying them to actual SAR data.

Chapter 4, "Parameter Estimation," is the perfect complement for Chapter 3 and for Chapter 5, "Applications". Reading Chapter 5 was a pleasure (as I am sure it will be for anyone with experience in image denoising). The classical filters (the mean, the median) commonly used to denoise natural images are explained, coded, and applied to SAR images. The Lee filter must be in any book dealing with SAR and speckle, and also it must be well addressed, as is the case in this book. Nonlocal means filters, also widely used for filtering natural images, are also discussed in Chapter 5. What it has been a pleasant surprise to find was the enhanced versions. Such improvement comes by introducing the Gamma law (so much discussed in the book) into the standard nonlocal mean schema. A quick look at stochastic distances enriches the analysis. Some classical classification methods (nearest neighbor, knn, K-means) are briefly explained and applied to simulated and synthetic SAR data.

At last, some advanced topics are gathered in Chapter 6, "Advanced Topics."

I did not address, intentionally, Chapter 2, "Elements of Data Analysis and Image Processing with R" in the above summary just to comment on it jointly with Chapter 7, "Reproducibility and Replicability." I said before that this is not just another SAR book, which I am sure it is not, even when not including Chapters 2 and 7, but certainly they will dissolve any doubt. The use of `R` (free and very powerful software) seems to be the correct decision: it makes it possible for any researcher to run the provided codes (all codes for all examples within the book are available for readers). This, of course, reinforces the authors' purpose when writing the book: make every result reproducible. The benefits of reproducible research are limitless, as they are detailed in Chapter 7. For even wider reproducibility, some parts of the codes are available in `R` and Matlab. The reader is referred to the website www.wiley.com/go/frery/sarimageanalysis for such contents.

To whom do I recommend this book? First, to any student or researcher interested in acquiring a sound knowledge of all mathematics involved in the SAR data. I also recommend this book as a textbook for an advanced introductory course to Image Processing for SAR. Finally, this book will also be of great interest to anyone out of the SAR scope who aims to understand better how mathematics makes complex image processing tasks easier. I tried to be as objective as possible when writing this Foreword: you, the true objective reader, will decide if this book helps you to grow in your understanding of this exciting topic.

Las Palmas de Gran Canaria, Spain *Luis Alvarez*
September, 2021

Foreword by Nelson D. A. Mascarenhas

Synthetic Aperture Radar (SAR) is one of the most important sources of images for remote sensing. Yet, it presents fundamental challenges for its interpretation. It is a great pleasure to accept the honorable and kind invitation by the authors of the book *SAR Image Analysis — A Computational Statistics Approach.*

I was one of the advisors of the doctoral dissertation by one of the authors, Alejandro C. Frery, thirty years ago and I have been following his brilliant career. He proposed extremely important tools for SAR image analysis, such as the G family of statistical distributions and the use of stochastic distances for the Information Theory based SAR image segmentation, among others.

The book covers all aspects of SAR in a very readable way, from data acquisition, data analysis, and image processing using the R language, going through the statistical models for intensity data, parameter estimation, applications, advanced topics, and concluding with the very important aspect of reproducibility and replicability.

The work relies on very formal statistical basis but, at the same time, it covers practical questions, including R code for the readers.

Concluding, I believe that this is a very welcome addition to the literature on SAR and one of the most complete sources both for the researchers, as well as the student who begins to explore the exciting area of SAR image processing and analysis.

São Paulo, SP, Brazil
September, 2021

Nelson D. A. Mascarenhas

Foreword by Paolo Gamba

The book in your hands is primarily a textbook, explaining in detail the topic of Synthetic Aperture Radar (SAR) statistical data processing. It contains theory, examples, and software to reproduce the examples and start working on new data sets. It presents a comprehensive state-of-the-art approach to the statistical analysis of SAR data. It is based on the work by the authors, who are well-known experts in the field, and experienced researchers. I have known two of the three authors of this book for many years. I have followed some of the live and online lectures by Alejandro Frery on many occasions, and I have been consistently amazed by their clarity and accuracy. I believe this book is no exception. It teaches you some R programming, too, which is a very nice and useful addition.

The book you are going to read is very accurately designed to meet requests from by readers interested in the specific topic of statistical data analysis applied to SAR and is an excellent introductory text for researchers and PhD students working on SAR data processing. It can be used for MSc courses, too, although in that case I would use more of the practical part (software and applications), skipping some of the details of the mathematics which is the basis of the statistical analysis.

Moreover, this very book fills a gap in technical literature and in my personal library (maybe in yours, too). As a matter of fact, there have been only a few books so far on the topic of SAR image analysis, and there is the need to have a publication describing the newest approaches in statistical models for SAR data, and the estimation techniques that may be used to extract the model parameters and validate statistical assumption with respect to data, procedures, and techniques applied to SAR data. The part about statistical models is very accurate, and clear and complete as I have expected. I am grateful to the authors for their work to put together all this information.

Last but not least, the second part of this book is in line with my engineering background, which makes me very keen to look for applications. I appreciate the way this book starts from the basics of SAR and moves to SAR statistical analysis, ending with interesting examples and applications. Statistical analysis of SAR images is not just mathematical stuff.

Therefore, you can be certain that this book will be useful to you, your curiosity, and your research. I hope you will enjoy reading it as much as I did.

Full Professor of Telecommunications and Remote Sensing
University of Pavia, Italy

Paolo Gamba, PhD

Foreword by Xiangrong Zhang

The development of remote sensing and its increasing societal and economic impacts have motivated in-depth research and wide applications of remote sensing technologies. In recent years, Synthetic Aperture Radar (SAR) imaging has demonstrated its strengths in earth observation as it can perform all-weather, high-resolution, large-area detection of ground objects. As a result, the analysis and applications of SAR images are one of the most critical frontiers in remote sensing.

SAR image analysis has gradually become indispensable for many disciplines and governmental offices to assist scientific research and solve challenging practical problems. As a specialist in this field, Professor Alejandro C. Frery has compiled his experience in this book, namely *SAR Image Analysis — A Computational Statistics Approach*. I am honored to write this preface.

I have known the authors Prof. Frery and Dr. Jie Wu, for many years. Dr. Jie Wu is not only a modest and meticulous scholar but also full of innovation. He specializes in SAR image processing with a statistics approach. Prof. Frery is a well-recognized academic scholar and a respectful professional. I have invited Prof. Frery to visit Xidian University several times. He delivered a course to post-graduate students during his visits, called "Operational Statistics for SAR Imagery." He taught the students the theory and practical work of SAR image analysis from elementary to advanced levels. All the students in the class were impressed by his theoretical knowledge and implementation skills. We all expect Prof. Frery to transfer his technical knowledge and experience to a book, and as we expected, the book now comes.

The authors of the book have paid tremendous efforts to its successful completion, where each chapter results from decades of research experience of the authors in this field. Statistical modeling of SAR images is one of the fundamental problems in SAR image interpretation. The book interfaces with several areas such as pattern recognition, image processing, signal analysis, probability theory, and electromagnetic analysis. This in itself is extremely demanding to the authors' theoretical knowledge in these areas. In addition, the book also presents exemplar codes corresponding to the theoretical chapters. From this perspective, the reader will appreciate the depth of the authors' research.

In a nutshell, this book mainly covers four main themes. Although this book focuses on the dedicated signal processing algorithms as the core of SAR image processing, Chapter 1, entitled "Data Acquisition," provides an overview of the performance of SAR and also includes an explanation and discussion about conventional radars. Secondly, from the third to the sixth chapters, this book introduces related disciplines to extend the learning on SAR image processing. Specifically, in Chapter 3, "Intensity SAR Data and the Multiplicative Model," the book's authors introduce fundamental theories and deduce the basic laws of SAR data before explaining the multiplicative models. In Chapter 4, "Parameter Estimation," the authors continue to discuss challenging problems but focus on modeling the acquired SAR data. These techniques form the basis of SAR image processing. Chapter 5, "Applications," the book focuses on integrating technologies of digital image processing, pattern recognition, and other disciplines. In particular, they introduce three classic filtering methods to deal with possible noise in SAR images. Meanwhile, several classification and clustering models have also been applied to SAR data. In Chapter 6, "Advanced Topics," the authors discuss commonly used image quality and algorithmic robustness evaluation metrics. Thirdly, this book focuses on the corresponding R language in Chapter 2, and many examples are also included throughout the book. Finally, in Chapter 7, "Reproducibility and Replicability," the authors summarize the development of multiple disciplines and derive the general law of engaging in scientific research. This is of great help to the researchers who even have been in the field for many years.

I want to thank the authors for their contribution to the materials included in this book. I am sure that this book is an invaluable reference manual to the researchers working in SAR analysis. For the reader who has learned the basic concepts of SAR, this book provides advanced and practical code examples to work with. This book can be used as a SAR textbook for university under- and post-graduate students.

Full Professor of School of Artificial Intelligence
Xidian University, China

Xiangrong Zhang, PhD

Preface

This book summarizes my experience in the field of statistical SAR data analysis. I started this journey back in 1991, during my second year as a Ph.D. student under the supervision of Profs. Nelson Mascarenhas and Oscar H. Bustos, at the *Instituto Nacional de Pesquisas Espaciais*, São José dos Campos, Brazil.

This book's seed was a twenty-hour course on statistical tools for SAR imagery analysis at CONAE – *Comisión Nacional de Actividades Espaciales*, Argentina. I wanted to make a course that would leave a positive trace: students would have the chance to learn and practice, and in order to do that I decided to use and share all the contents in `RMarkdown`. It was a nice experience, but I felt limitations (mostly mine) with such a platform. When I was later invited to produce material for a distance education course, my first challenge was deciding the format. After pondering a couple of possibilities, I opted for a classical book.

Selected material from this book has served for courses in Argentina, Brazil, Chile, China, Colombia, Costa Rica, India, Italy, Israel, and Nigeria. I am grateful for the students that attended those courses for their positive and constructive feedback.

One of the books that exerted the strongest influence in my career was *The Visual Display of Quantitative Information* (Tufte, 2001). I have tried to follow his guidelines while producing the graphical material, which was constrained to black and white by the publisher.

Writing a book is a frightening task, but I have learned a few lessons from the past, among them:

- Reading with the eyes of the public helps with staying focused.
- Using LaTeX and BibTeX is absolutely mandatory for quality results.
- Control version (Git in this case) is a safety net.

So here we go with this book. I hope it will be helpful to those that have the curiosity to browse through its pages.

Wellington, New Zealand *Alejandro C. Frery*
September, 2021

Due to the coherent imaging scheme used in the SAR system, random speckle is inescapable in SAR imagery. Therefore, using a statistics viewpoint is essential for the analysis and processing of SAR images.

When I entered graduate school, SAR image despeckling became my first topic for study. After reading too many relative papers, I found that statistics was an excellent tool for the issue, especially for the description of randomness. Then, I decided to rebuild my foundation in probability and statistics step by step. Due to the gap between the application and fundamental knowledge, I put more energy into searching for the relative message to bridge these two components. Moreover, with the deepening of research on SAR image processing, I found that almost all processes of SAR image need statistics.

In this book, we try to fill the gap and give all readers the direct link between the knowledge from the book and the detailed application, especially from a statistics viewpoint. Therefore, I hope this book will enlight all readers who want to enter the territory of SAR image processing.

Last but not least, it is my honor to work with Prof. Alejandro and Prof. Luis on the writing of this book, especially for the discussion about the content of the book. It is a great time to learn from these two excellent scholars.

Xi'an, China *Jie Wu*
September, 2021

I've been researching image processing and remote sensing (SAR and PolSAR) data for some time. In those years, I had the privilege to work with outstanding scholars and shared motivations for pushing ahead research a little.

Additionally, my experience as an Editor, author, and reviewer, keeps me in close contact with the state-of-the art methods and techniques and most of *all abouts* dealing with the challenging area of image processing.

It has been my priority when co-authoring this book, to introduce and discuss all topics within this volume in a flat but rigorous way, always with the reader in my mind as being *in front of me*.

I do believe in books, in learning from academic books. It is my honest hope that readers of this one would find every concept academically explained. My aim has been always to discussing all from the very beginning till the end, and also, providing the necessary references to enhance the knowledge. As a reader of academic books, I have always been pretty concerned that some details are missing that leads to a full understanding of what is discussed not being achieved. I hope that did not happen with this book you are reading!

Las Palmas de Gran Canaria, Spain *Luis Gomez*
September, 2021

Acknowledgments

It is impossible to make a complete list of those that contributed to this book; this work is, as I am, a convex combination of randomness and choices. I would like to express my deepest gratitude to my advisors, mentors, and role models: Professors Oscar H. Bustos, Nelson D. A. Mascarenhas, and Héctor Allende O. They taught me much more than technical things. My friends Corina da Costa Freitas and Sidnei J. S. Sant'anna are loyal companions since the beginning of this journey. My hosts in Lanzhou and in Xi'an, Profs. Xin Li and Xiangrong Zhang, encouraged me to work in this project, and provided ideal conditions for enjoying life while doing it. My friends from UFPE (Universidade Federal de Pernambuco, Recife), from LaCCAN *Laboratório de Computação Científica e Análise Numérica*, Universidad de Buenos Aires, Instituto Tecnológico de Buenos Aires, Universidad Nacional del Nordeste, Universidad Nacional de Río Cuarto, and Universidad Nacional General Sarmiento, and my students all collaborated with fruitful discussions and emotional support.

A. C. F.

The deep communication between Alejandro and me began during the summer holiday in 2019. As Alejandro is an excellent scholar in statistics applied for SAR image processing, I asked him too many questions about the usage of different statistic distributions. He tried his best to make a comprehensive explanation, even for formula derivation by hand. It was a great time that I received about the way for research in this area. At that time, Alejandro proposed I participate in this book. It is a great honor and pleasure for me. Thanks very much for Alejandro.

Given the invitation from Prof. Xiangrong Zhang to Alejandro for the summer school curriculum at Xidian University, I had the chance to communicate with Alejandro face to face. I want to show my great thanks to Prof. Xiangrong Zhang. She provides me with so much help in scholarly research.

Meanwhile, I want to show my thanks to Luis, who is the co-author of this book. He always gave me practical recommendations when I met a problem at work. It is an impressive time to work with him.

Last but not least, I want to thanks my tutor, my family, and my friends. They always give me helps and supports, enlightening the direction for me when I enter into puzzlement.

J. W.

When Alejandro invited me to participate in this book, the first thought that came to my mind was, what a tremendous responsibility! but also, what a great honor and pleasure! Thanks Alejandro!

My greatest debt, as always! is to my family, my wonderful partner, Stella, and our so lovely kids, Gabriel (*Gaby*) and Martina (*Martinita*). As the Ernesto Lecuona's old *bolero* sings: *siempre en mi corazón*, (*always in my heart*).

L. G.

Acronyms

AISAR	airborne synthetic aperture radar
ESAR	experimental synthetic aperture radar
HH	horizontal/horizontal polarization mode
HV	horizontal/vertical polarization mode
IQR	inter-quartile range
MAP	maximum a posteriori
ML	maximum likelihood
MSE	mean square error
NLM	nonlocal Means
SAR	synthetic aperture radar
SEIF	stylized empirical influence functions
VH	vertical/horizontal polarization mode
VV	vertical/vertical polarization mode

Introduction

The book describes, in a practical manner, how to use statistics to extract information from SAR imagery. It covers models, supplies data and code, and discusses theoretical aspects, which are relevant to practitioners. It provides a unified vision of the field.

This book does not cover complex signal processing methods and hardware issues involved in SAR, although a brief introduction is provided in Chapter 1, "Data Acquisition". As practical users of SAR data, we deal with final products: SAR images.

In this chapter, we provide a brief introduction to statistical analysis of SAR (Synthetic Aperture Radar) images. Then, the organization of the book is presented.

I.1 SAR

Synthetic Aperture Radar (SAR) sensors have a prominent role in remote sensing with microwaves. They can provide images with high resolution (up to centimeters on the ground), information about de dielectric and textural properties of the target, they operate without need of sunlight, and are almost immune to adverse visibility conditions (clouds, rain, fog, etc.). The capabilities of SAR to monitoring the Earth's surface are unique, providing remote sensing data at low-high spatial resolution 24 hours, 365 days a year. Such capabilities are not offered by other systems. That is possible because SAR systems use active sensors to illuminate the area under observation (as a difference with optical sensors, that use passive sensors and also are affected by the atmospheric conditions). Other remote sensing systems that use active sensors, for instance LIDAR (Light Detection and Ranging o Laser Imaging Detection and Ranging) are also very useful, but SAR systems employ sophisticated signal processing techniques that allow to getting precious information from the scanned area and targets.

The first SAR system on a space mission was aboard of the Seasat satellite launched in 1978. This first SAR observational system scanned along 126 million square kilometers of the Earth's surface from an altitude of 800 km. The captured data resolution was 25 m. Modern SAR systems provide data at resolutions of 20 cm.

Therefore, SAR systems provide invaluable information for capturing data in a global scale. SAR data provides information for applications such as monitoring agricultural uses, observing landslide and changes in the land and uses, tracking oil spills, analysis of temperatures changes (land, water, oceans,...), surveillance, military applications. Its role in the studio of climate change consequences, among other challenging problems of interest for humankind, makes SAR systems irreplaceable.

I.2 Statistics for SAR

The data from SAR sensors suffer from an interference pattern called speckle that requires specific models and tools. Analyzing SAR data is both challenging, because usual models should not be applied, and rewarding, as their statistical properties reveal relevant features of the scene.

The area of statistical modeling of speckle began in the middle of the past century, with the pioneering works by N. R. Goodman who studied optical speckle. His work, in a nutshell, proved that the simplest intensity model is the Exponential law. A second milestone in the field was set by E. Jakeman and R. J. A. Tough, in 1987, when they derived the K distribution. These models share a common property: they are parametric, and the estimation of the parameters plays a central role in the extraction of information. Models commonly used in SAR image processing, and state-of-the art models and statistical techniques constitute the central core of this book.

In a different line of research, Deep Learning is gaining space in making these tasks more automatic. Nevertheless, the classical approach that we adopt in this book is likely to continue being the touchstone of the area, because of the connection of the statistical models with the physical properties of the scene.

Essentially, the reader will learn specific concepts of non-Gaussian data analysis, as applied to speckled data. This departure from the classical approach to data analysis is a valuable difference from standard textbooks and courses, as it prepares both the researcher and the practitioner to face practical challenges that appear in real-world applications.

I.3 The Book

This book brings a fresh view to this field, by gathering the theoretical properties of adequate models, estimators, interpretation, data visualization, and advanced techniques, along with data and code snippets. Many of those contents are scattered in specialized literature, aimed solely at researchers and practitioners of microwave remote sensing. This work, besides making a unified presentation, offers a wealth of information in such a way that can be used in Data Analysis and Statistics courses, as valuable examples in which classical techniques should not be applied.

The knowledge and tools conveyed by the book are scattered in the scientific literature and, to the best of our understanding, there are only a few examples of works that provide a comprehensive view of the theory, of the algorithms, of their implementation, of the application of the tools and, ultimately, of the assimilation of the information extracted from the data.

The targeted audience consists of both researchers and students in Remote Sensing, Data Analysis, and Statistics. The book can be both seen as at intermediate level and advanced. Practitioners may skip the theoretical parts, and jump to the applications, while advanced students and researchers will benefit from the in-depth theoretical contents. The book includes exercises and research topics.

The organization of this book is as follows,

Chapter 1, *Data Acquisition*, provides an introduction to SAR systems and how data are acquired, whereas the main parameters related to SAR systems such as azimuth resolution and range resolution and the main acquisition modes are introduced. A simple introduction to radar systems is included to better deal with the interaction of the radar emitted and transmitted signals with targets. We pay special attention to signal backscatter due to its relation to speckle. A brief description of the currently deployed satellite providing SAR data and useful tools to deal with freely accessible SAR data is also provided. Therefore, Chapter 1 is just an Introduction of SAR avoiding getting into too technical complex details.

In the Chapter 2, *Elements of Data Analysis and Image Processing with R*, some fundamental knowledge and tools about the data analysis will be discussed, as the statistical properties of SAR data are extremely important for SAR image processing. Additionally, due to the spatial property of SAR images, some operations about image processing are also introduced. To make a further understanding of the content of this chapter, corresponding codes based on `R` are given. The concepts covered in Chapter 2 will ease readers of this book to easily execute all R scripts available at www.wiley.com/go/frery/sarimageanalysis.

The material in Chapter 3, *Intensity SAR Data and the Multiplicative Model*, provides the reader's first exposure to the statistical modelling of SAR data. In this chapter, the basic properties of SAR data, starting from the complex scattering vector and then reaching the Exponential and Gamma distributions are derived. With this, what many authors call *fully developed speckle*, or *speckle for textureless targets* is covered. The models discussed will be generalized later for other situations of great interest (both theoretical and practical).

The discussion in Chapter 4, *Parameter Estimation*, brings out the key role played by estimation in the statistical modelling of data. At this point of the book, we have data and models: We will explore ways of using the former to make inferences about the latter.

With all this sound theoretical background, it is time to use them in practical SAR image processing matters.

Chapter 5, *Applications*, is devoted to applying what has been learned in previous chapters. Despeckling filters based on statistical methods for SAR imagery have been chosen to put into practice the statistical models studied due to their relevance in the analysis and interpretation of SAR images. Image classification has been also addressed through the standard (classical) elemental machine learning methods. The mean, median, and the Lee filters must appear in SAR books dealing with speckle. Then, the Maximum a posteriori (MAP) filter and, the non-local approach are discussed (the original and the state-of-the art statistical non-local filters using stochastic distances and hypothesis test). These despeckling filters are commonly used in SAR despeckling, providing excellent results. For each filter, a brief introduction is first addressed, and then, some applications for both, simulated and actual SAR data are given. Many examples are shown and also, all the codes are available at www.wiley.com/go/frery/sarimageanalysis.

In Chapter 6, *Advanced Topics*, we get back to theoretical issues of relevance in SAR and in statistical modelling. The first topic deals with the assessment of despeckling filters, which is not a trivial issue. It resembles a multifaceted problem where many aspects must be considered. A set of well-established image-quality indices (*metrics*) are first introduced. Then, new advanced metrics for assessing a filtering operation for the challenging case of no-reference image available are introduced. This chapter finishes with the exposition of a fundamental statistical topic: robust inference. Such concept should always be part of any statistical analysis, but it is also of paramount importance in image processing. The aim is to clarify the concept of statistical robustness in the context of SAR data analysis by providing examples of what is needed to fairly use the term "robust" when attached to inference or to a procedure that depends on estimation. We provide examples and hints to develop robust inference techniques.

The final chapter of the book, Chapter 7, *Reproducibility and Replicability*, provides a discussion on reproducibility and replicability (of methods, algorithms, research,...). Both terms are frequently confused even among researchers, therefore, at first, it is important to provide a clear definition for each word. Then, the implications of such relevant concepts in Remote Sensing in general, and in particular to the purposes aimed in this book, are discussed. In Section 7.4, a list of recommendations and suggestions to adhere to the good practices for making good research in SAR is drawn out.

More about reproducibility and replicability is discussed below.

Most chapters of this book include a collection of problems, always closely related to the theoretical and practical explanations provided in each chapter. We strongly encourage the reader to solve them to take full benefit of this book. All codes are also available as it is detailed in Section I.4.

I.4 Commitment to Reproducibility and Replicability

Discuss models, speckle, statistical distributions, etc. with clarity, soundly, in a dynamic and educational way. It was one of the priorities of the Authors of this book. But in the mind of all, reproducibility and replicability were as important as the mathematical notation used.

It is important to notice that reproducibility and replicability should permeate the whole scientific life. As noted by Wilson and Botham (2021), addressing these issues enhances your proposal by adding rigor and credibility.

The reader will find it easy to reproduce and to replicate all results contained in this book. In doing that, it is our desire the reader will adhere to the benefits of allowing anyone to reproduce and to replicate his/her work and also to follow the "good science" practices.

Being `R` the computational platform of choice, this text will reach a large statistical community. Most of the codes are in `R` but, we are aware also that a large part of the SAR community uses other tools. As it is not possible to satisfy all, some codes are in Matlab. In some cases, (non-local filters) both versions, `R` and Matlab, are included. Codes are commented and they are easy to follow. Although they are computationally efficient, sure they can be improved.

About the Companion Website

This book is accompanied by a companion website

www.wiley.com/go/frery/sarimageanalysis

The website includes data and code.

1

Data Acquisition

In this chapter, we focus on Synthetic Aperture Radar (SAR) data. We first provide an introduction to SAR systems and detail how data are acquired, whereas the main parameters related to SAR systems such as azimuth resolution and range resolution and the main acquisition modes are introduced. A simple introduction to radar systems is included to better deal with the interaction of the radar emitted and transmitted signals with targets. We pay special attention to signal backscatter due to its relation to speckle. A brief description of currently deployed satellite providing SAR data and useful tools to deal with freely accessible SAR data is also provided.

1.1 Introduction

Figure 1.1 summarizes this chapter in the context of this book: a sensor (Sensor) mounted in a spacecraft (a satellite, an aircraft, or a ground-based platform) captures signals *remotely* over the Earth's surface that is conveniently processed (Signal Processing) and transformed into data (Data). Users build models to deal with the data to suit their needs. Due to the inherent random nature of data (see Chapter 3), statistical modeling is of maximum interest and it seems to be the first approach to handling the data. *Speckle noise* (or just *speckle*) is a particular kind of *noise* that largely corrupts the SAR data, making it difficult to interpret acquired images. How speckle is related to image pixels is also accounted in this chapter. Once data have been properly processed, despeckling is a major issue (see Sections 5.1 and 5.2.2), it is when applications enter into play. From the many possible uses, image classification (see Sections 5.5, 5.6, 5.7, and 5.8) and image segmentation are the most common image processing tasks.

Remote observation of the Earth's surface can be done by using passive or active sensors (see Figure 1.2). A passive sensor (shown in (a)) does not emit any signal to the Earth's surface to collect data: it just receives the signal. Such signal is emitted by the Earth's surface in the form of radiation or is due to the reflected sunlight, or both phenomena. Cameras on satellites or on aircrafts are examples of passive sensors. However, active sensors (shown in (b)) first emit a signal that interacts with the Earth's surface and the reflected signal is detected by the sensor. This chapter (and this book), only deals with active sensors for the reason that acquisition system for collecting data discussed in this book uses them.

SAR Image Analysis — A Computational Statistics Approach: With R Code, Data, and Applications, First Edition.
Alejandro C. Frery, Jie Wu, and Luis Gomez.

Companion website: www.wiley.com/go/frery/sarimageanalysis

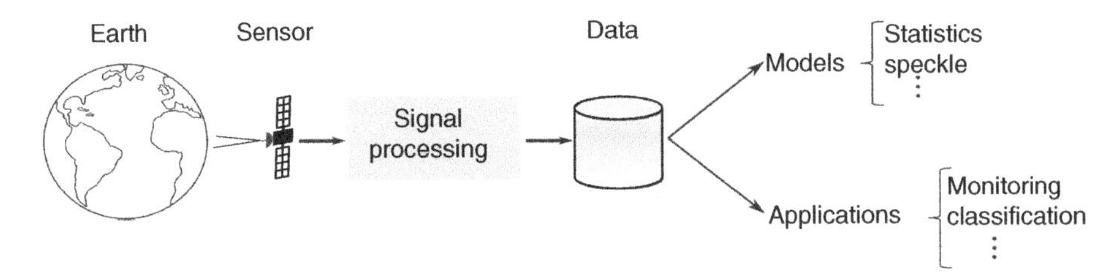

Figure 1.1 Remote sensing data acquisition and data processing.

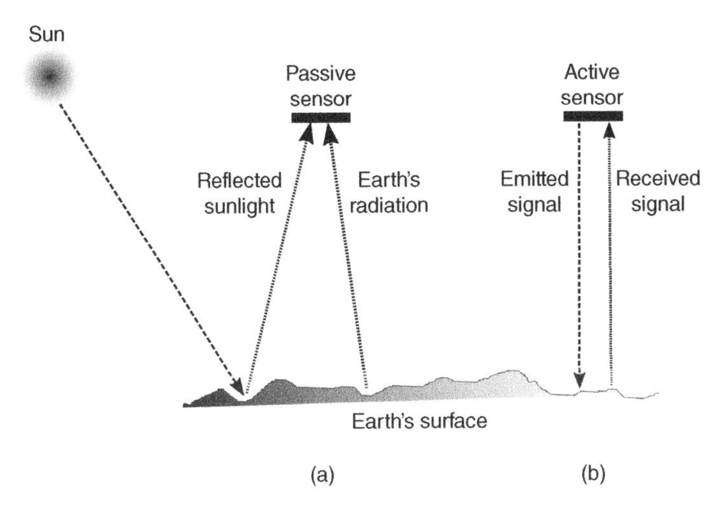

Figure 1.2 Passive sensor (a) and active sensor (b).

1.2 SAR

Synthetic Aperture Radar (SAR) sensors have a prominent role in remote sensing with microwaves. They can provide images with high resolution (up to centimeters on the ground), information about de dielectric and textural properties of the target, they operate without the need of sunlight, and are almost immune to adverse visibility conditions (clouds, rain, fog, etc.) Therefore, SAR systems are capable of providing high valuable data to users 24/7 all year round. That is possible because SAR systems use active sensors to illuminate the area under observation (as a difference with optical sensors, that use passive sensors and also are affected by atmospheric conditions, such as clouds, day/night cycle, rain, etc.) Before addressing the standard configuration of a SAR system, to better understand them first, we explain how a radar works.

1.2.1 The Radar

Radar (Radio Detection And Ranging) systems were developed for military use during the second World War for surveillance purposes. Radar rapidly motivated the studio of signal processing techniques and, also, encountered many new areas of applications. In the military area, it includes air-defense systems and antimissile systems, among others, whereas civil applications span from astronomy (radio-astronomy) to aircraft navigation (fight control

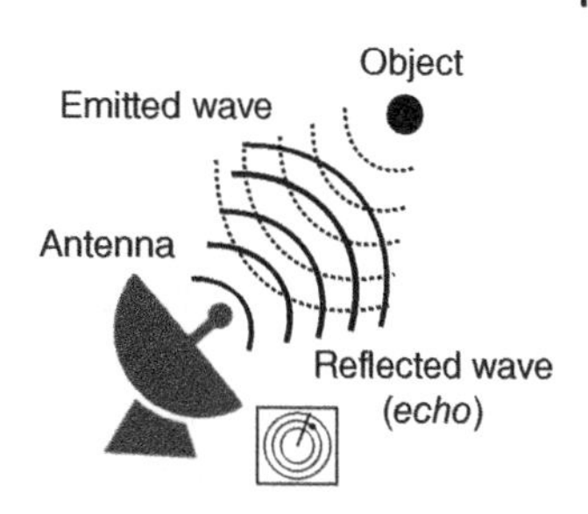

Figure 1.3 Simplified radar emitting-receiving operation.

systems, altimeter measures, etc.) In the particular area of remote sensing, radar was first used for monitoring precipitations and it soon stemmed as a very useful tool for geological observations.

As it is generally known, radar operates by transmitting an electromagnetic signal (*the radar signal*) over an area (*illuminated area*) and getting information from the reflected signal (received signal) due to the illuminated object (target). So, the radar uses an active device (the emitting antenna) that, in most cases, is also the sensor that captures the reflected signal (see Figure 1.3).

Obtaining information from the received signal is not a trivial issue, and it requires deep knowledge of electromagnetic propagation, signal scattering, and signal processing. Fortunately, all this knowledge is available at present, after decades of intensive research over the past century. The generation of the signal to transmit as well as the amplification of the received signal (usually very weak) and other signal conditioning (filtering, quantization, analog-digital, digital-analog conversion, etc.) requires also the use of specific electronic systems also largely studied in the past. The transmitted pulse used is a *chirp* (see Figure 1.4): a signal that contains many frequencies, increasing (chirp-up), as in the case shown in the figure, or decreasing (chirp-down). The benefits of the chirp signal over a simple single-frequency pulse are:

- it provides better resolution,
- the area covered (illuminated) by the signal is enhanced,
- it is less affected by noise,
- the circuits used to generate such signal are relatively simple,
- the system requires less power.

In Figure 1.4 it is also shown the basic radar operation principle: each target reflects (scatters) the received signal in many directions, and the reflected signals toward the transmitter are the desirable ones that make radar work. As also illustrated in this figure, the reflected signal (echo signal) is a variation of the emitted one. Such variations in amplitude, wave

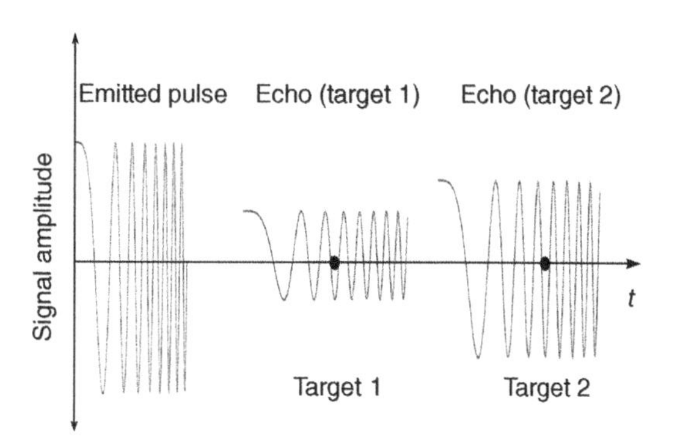

Figure 1.4 Simplified radar emitting-receiving operation using a chirp-up signal (emitted pulse).

shape, and frequency, are caused mainly by the physical properties of the targets. In this way, the return does not only provide information about the distance from the emitter (basic radar operation), but also valuable physical information (advanced systems such as SAR). Although the basic principle of radar operation is indeed simple to understand, radars are complex systems, and a deep knowledge of signal processing is required to fully understand how the echoes are finally converted into useful information. However, it is not difficult to understand that, as the emitter signal is totally known (the chirp signal), by comparing the emitter signal with the received one, targets placement can be easily estimated. This operation is done through the well-known matched filter that, in essence, performs the correlation between the emitted and the received echos.

After this brief introduction to the basic radar principles of operation, SAR systems are explained in subsection 1.2.2.

1.2.2 What is SAR?

An extensive set of specific mathematical and signal tools for radar applications were available early in the mid-fifties of the past century. They paved the way for new systems to come. SAR (Synthetic Aperture Radar) was invented by Carl A. Wiley in 1951 and, after solving many technical problems (especially ones related to keeping moving trajectory stable), the first acquired SAR image was obtained in 1957. The first SAR system on a space mission was aboard of the Seasat satellite launched in 1978. This first SAR observational system scanned along 126 million square kilometers of the Earth's surface from an altitude of 800km. The captured data resolution was 25 m (both in range and azimuth directions)[1]. It is worthy to mention that the NASA's space Shuttle carried an enhanced version of the Seasat satellite in 1981 collecting, in only three days, image data of about 10 million square kilometers of the Earth's Surface from an altitude of 245 km. Resolution of data was 40 m (both in range and azimuth directions). After that, most Shuttle missions incorporated enhanced SAR systems.

SAR systems are frequently placed on military and commercial aircrafts, reducing costs when compared to satellite deployments. A great advantage of airborne SAR systems is the greater flexibility in establishing data acquisition parameters such as the incidence angle and the resolution. Observation altitude is evidently lower so, the although the scanned area is also reduced, the resolution can be much better (https://ui.adsabs.harvard.edu/abs/2017EGUGA..1918019P/abstract). For a description of an airborne system see https://www.dlr.de/hr/en/desktopdefault.aspx/tabid-2326/3776_read-5691/. This airborne SAR system provides data with resolution of 20 cm in the azimuth direction and it is mounted on a Dornier DO228-212 aircraft. Boeing E-7 Wedgetail AWACS (Airborne Early Warning and Control), NATO (North Atlantic Treaty Organization) military planes include SAR systems (https://www.thedrive.com/the-war-zone/39451/top-air-force-general-in-the-pacific-wants-e-7-wedgetails-to-replace-e-3-radar-planes).

What is SAR? Among the many differences between a standard radar and a SAR, the most important are:

- SAR is not just a radar: it is an imaging radar system,
- SAR uses a sensor (or antenna) that is not fixed but it is moving as it emits signals and receives the echos.

1 These concepts are described in Section 1.2.4.

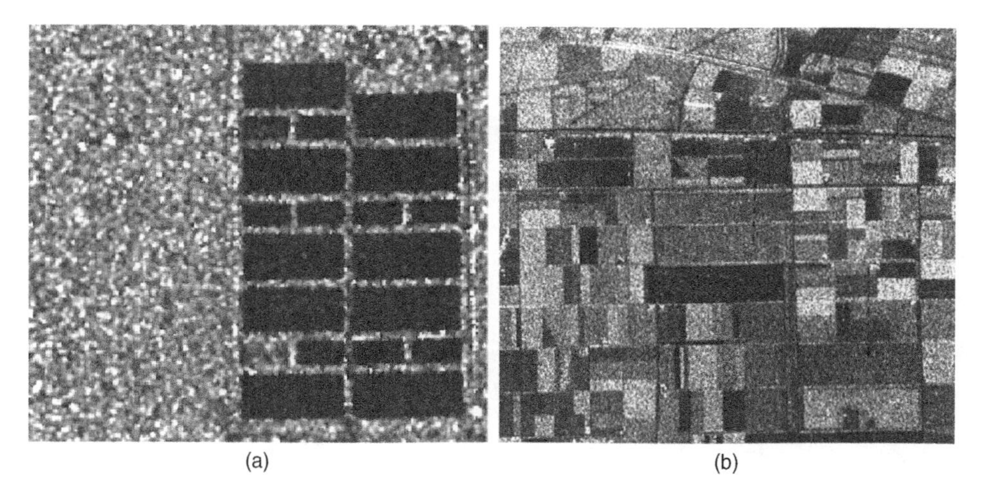

Figure 1.5 (a) An actual single-look image from the COSMO-SkyMed SAR sensor acquired on June 5, 2018 over the Amazon rainforest (black areas are fish ponds). The spatial resolution (in range and azimuth) is 3 m. (b) Airborne Synthetic Aperture Radar (AIRSAR) sub-image of Flevoland agricultural area in The Netherlands. It was obtained by the NASA / Jet Propulsion Laboratory AIRSAR platform in 1989 (azimuth resolution is 12.1 m and range resolution is 6 m). Source: NASA / Jet Propulsion Laboratory AIRSAR.

Therefore, SAR systems do not provide an *image* of the echos received, but, through the use of complex signal and image processing techniques, it obtains images (*photo-like*) of the illuminated area. It can be noticed that, although SAR images are somehow similar to natural images, they have a significant amount of noise (this will be discussed in Section 1.5). An example of SAR images (actual data) can be seen in Figure 1.5.

Figure 1.6 shows the same scene, the well-known San Francisco Bay area, as seen by an optical sensor and by a SAR. It is noted that, although both images are visually quite different, many targets are clearly recognizable in the SAR image (the ocean, the blocks of buildings, the forest and parks, the Golden Gate bridge, and the Alcatraz island are easily seen). Moreover, the Golden Gate is even more visible in the SAR image than in the optical one because metallic structures reflect the radar signal strongly. Note that the SAR image could have been acquired during the night or on a cloudy day.

As indicated above, the SAR antenna moves when emitting and receiving signals. In Figure 1.7 a simplified description the geometry of a SAR system is depicted. The sensor movement provides the high-resolution capabilities of the SAR, as well as it introduces complex calibration and stabilization challenges, fortunately, largely solved at present.

After this brief introduction to SAR, Section 1.2.3 focuses on particular details of actual SAR systems.

1.2.3 SAR Systems

As explained, a SAR sensor is an imaging radar system that emits pulses (electromagnetic waves) whose wavelengths are of the order of centimeters, ranging from about one meter to one millimeter corresponding to frequencies between 300 MHz to 300 GHz, respectively, as follows from,

$$f = \frac{c}{\lambda}, \tag{1.1}$$

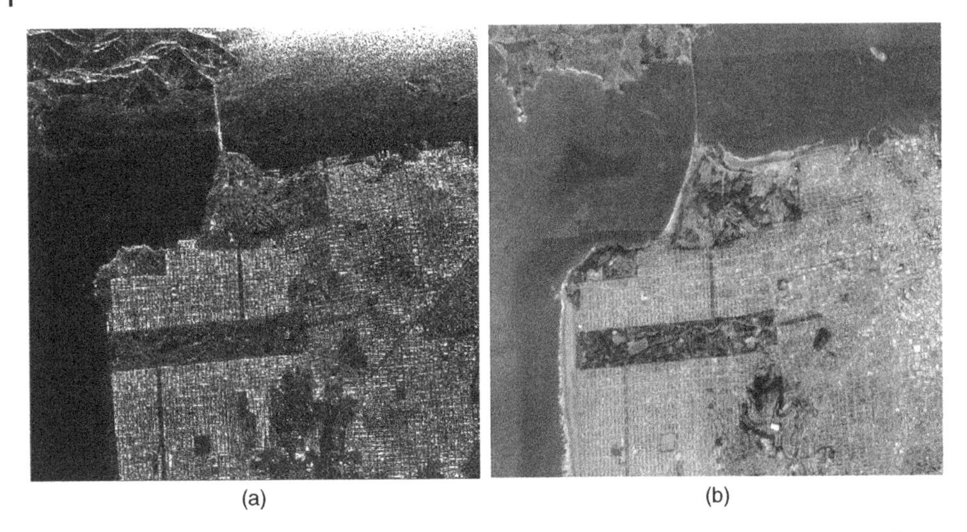

Figure 1.6 (a) NASA/JPL AIRSAR image for San Francisco Bay and the optical version, from Google Earth (b). The SAR image has spatial ground resolution of about 6.6 m in range and 9.3 m in azimuth. The images have not been registered (aligned) and some additional differences between them are due to geophysical corrections applied to the SAR data. Acquisition dates are not either the same.Source: (a) NASA/JPL AIRSAR and (b) Google Earth.

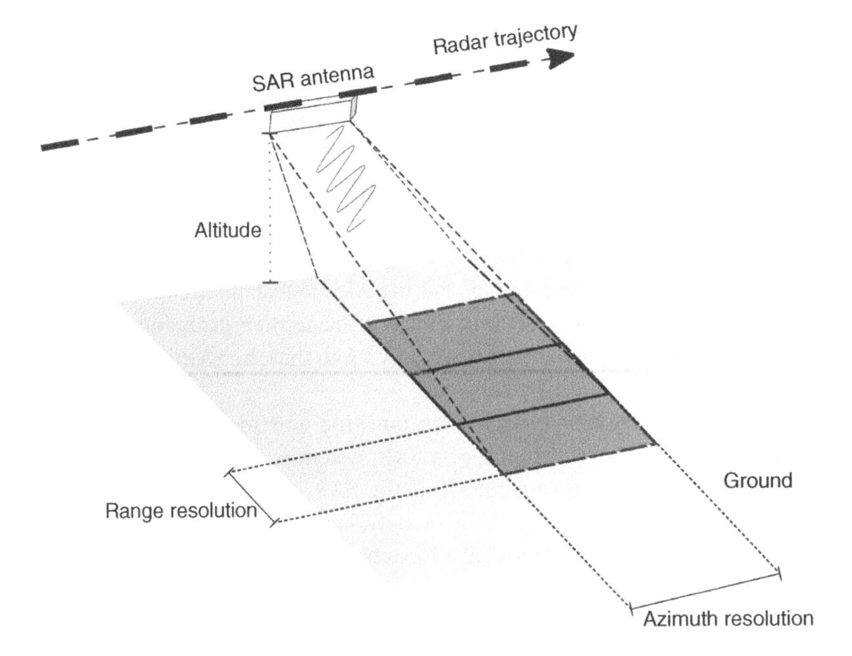

Figure 1.7 Synthetic aperture system (simplified description).

where c is the speed of the light[2] ($\approx 3 \times 10^8$ ms^{-1}) and λ is the wavelength of the signal (in meters). Figure 1.8 shows two waves with different wavelengths. The shorter the wavelength is, the more energy it has. Just to compare, the visible light ranges from 390 nm to 750 nm

2 All electromagnetic waves travel at speed c.

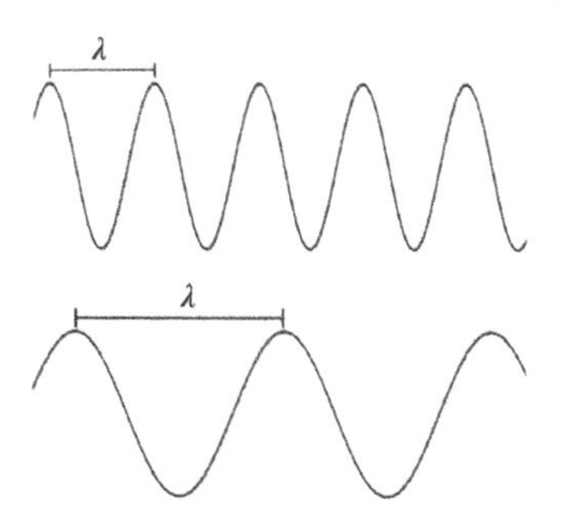

Figure 1.8 Two waves with different wavelengths (top wave has double frequency than the bottom wave).

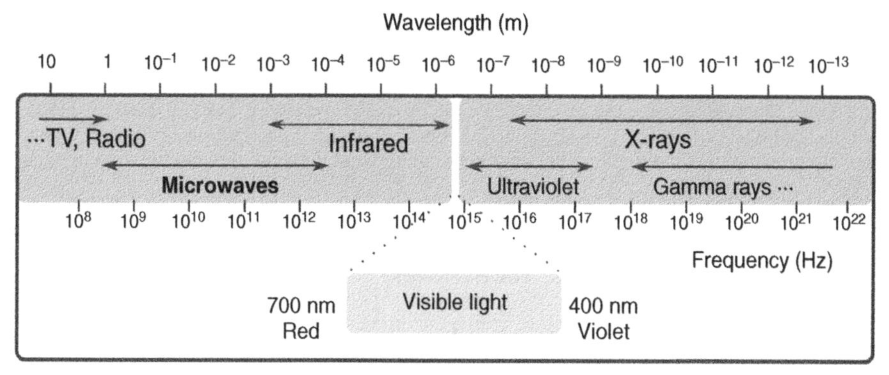

Figure 1.9 Electromagnetic spectrum.

(nanometers), whereas infrared radiation ranges from 750 nm to 1 nm (see Figure 1.9, where a region of interest of the electromagnetic spectrum is shown). Note also that as the wavelength diminishes (higher frequencies), the larger the resolution of the system is. Besides, microwave signals travel by line-of-sight (point-to-point).

Note also that, as mentioned above, what the sensor emits and receives is not a waveform like the one shown in Figure 1.8, but an electromagnetic signal like the one represented in Figure 1.10. This is of especial importance since both the electric and the magnetic components interact with the target providing, thus, much valuable (physical) information from the illuminated area. Additionally, the information collected by the sensor comes not only from the amplitude of the reflected signal but also from the change of its phase (this will be addressed below).

This book deals with SAR data in which the transmitted signal has a single polarization mode, unlike PolSAR (Polarimetric SAR) systems where several polarization modes participate. Nevertheless, it is convenient to have some knowledge of what signal polarization refers to.

Polarization (of a electromagnetic wave) refers to the orientation of the electric field during propagation respect to a reference. For convenience, the reference is chosen to be parallel to the Earth's surface on transmission, which is known as horizontal (H) polarization. For a general wave (see Figure 1.11), the horizontal polarization (H) indicates that the wave is oscillating within a horizontal plane as it propagates. The vertically polarized wave (V) oscillates within a vertical plane as it propagates. When the wave is not polarized, it vibrates in all planes orthogonal to the axis of propagation.

The electromagnetic wave, that is, the emitted and reflected radar signal, has two components (see Figure 1.10): the electric component (**E**), and the magnetic component (**B**), which

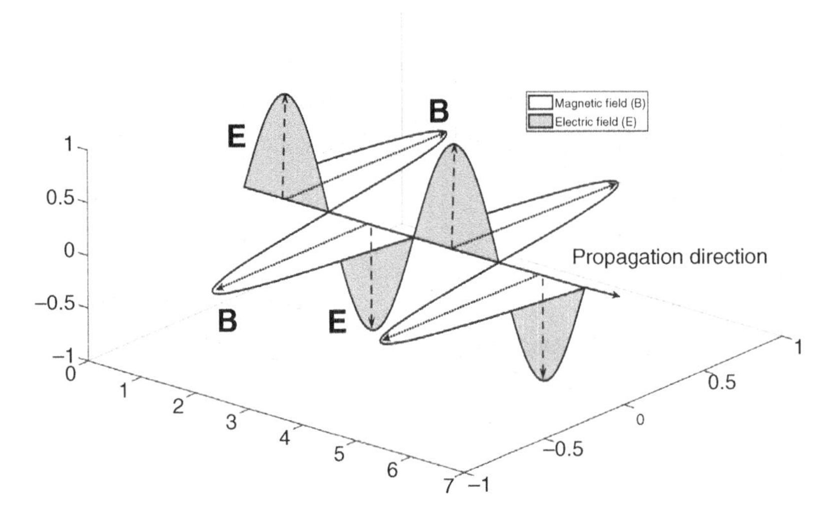

Figure 1.10 An electromagnetic wave (**E** and **B** vectors are orthogonal).

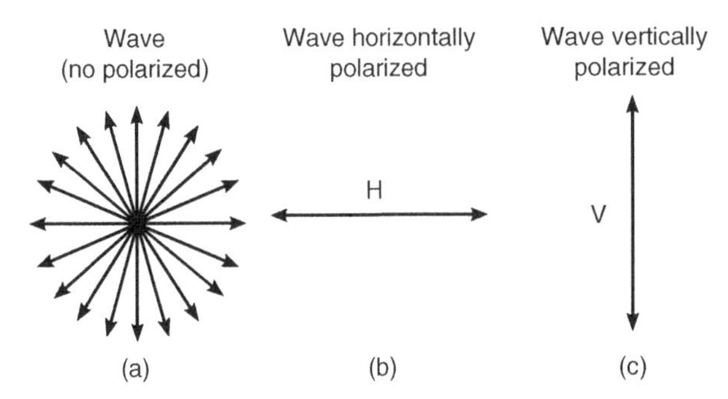

Figure 1.11 Not polarized wave (a), horizontal (b), and vertical polarization (c).

are orthogonal. In the radar's context, the polarization of a signal refers only to the orientation of the electric component, E, with respect to the emitter (antenna).

A radar antenna can be designed to send and receive electromagnetic waves with well-defined polarization. Two modes of radiation, horizontally (H) and vertically (V) polarized waves are possible. Figure 1.12 illustrates two sensors emitting an H wave (left) and a V wave toward the target. It is important to note that an emitted H wave, after being scattered by the target, may produce both H and V waves. The opposite is also true, an emitted V wave, after being scattered by the target, may produce H and V waves.

Therefore, the emitter can transmit either an H or a V wave. Additionally, the sensor detects only reflected H or V waves, depending on the established configuration (this can be dynamically modified). It is clear that the combination of emitted and detected waves produces four modes of radar operation:

Co-polarized modes: HH (signal emitted H, signal received H), and VV (signal emitted V, signal received V).

Cross-polarized modes: HV (signal emitted H, signal received V), and VH (signal emitted V, signal received H).

These modes of operation offer the user a vector related to each pixel of the image, so, better characterizing the target properties (target signature). As mentioned at the beginning of this

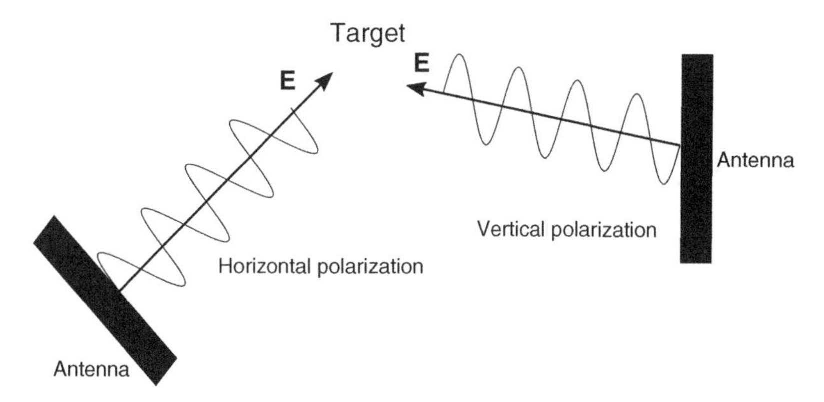

Figure 1.12 Illumination of a target by horizontal (a) and vertical (b) polarized waves.

section, PolSAR systems use such a rich combination of polarization modes providing much more information for targets than SAR systems that use a single-mode (single-polarization system or just "single-pol") that constitutes the object of this book.

The forthcoming description of modern deployed SAR systems requires the standard nomenclature related to the polarization modes:

- Single-polarization, or single-pol: transmits and receives a single polarization, typically the same direction, resulting in horizontal-horizontal (HH) or vertical-vertical (VV) imagery.
- Dual-polarization, or 'dual-pol': may transmit in one polarization but receives in two, resulting in either HH and HV, or VH and VV imagery.
- Fully polarimetric, polarimetric, or quad-pol systems alternate between transmitting H and V pulses, and receive both H and V, resulting in HH, HV, VH, and VV imagery. To operate in quad-pol mode the radar must pulse at twice the rate of a single- or dual-pol system since the transmit polarization has to alternate between H and V pulse by pulse.

Figure 1.13 summarizes the usual polarization modes.

Without entering much in explaining polarization mechanisms, which is beyond the scope of this book, it is known that:

- physical processes responsible for like-polarized HH or VV returns are related to quasi-specular surface reflections (calm ocean/water surfaces, i.e., without waves). Due to the specular reflection, the detected return signal is practically null, which is seen in the image as a black area (see the black large areas in the SAR image of San Francisco Bay image in Figure 1.6).
- HH and VV modes returns are stronger than cross-polarized (HV or VH) modes.
- Cross-polarized modes show less penetration effect (through canopy, sand, snow, or soil) than co-polarized modes.

Figure 1.14 shows the HH, HV, VV, and HH_VV (module) for an actual SAR image. As it can be seen, each polarization mode offers different information.

1.2.4 The Synthetic Antenna

The two first letters in SAR refer to *synthetic aperture* (or antenna) which is introduced in this section, as well as other important concepts related to the whole SAR system. To represent

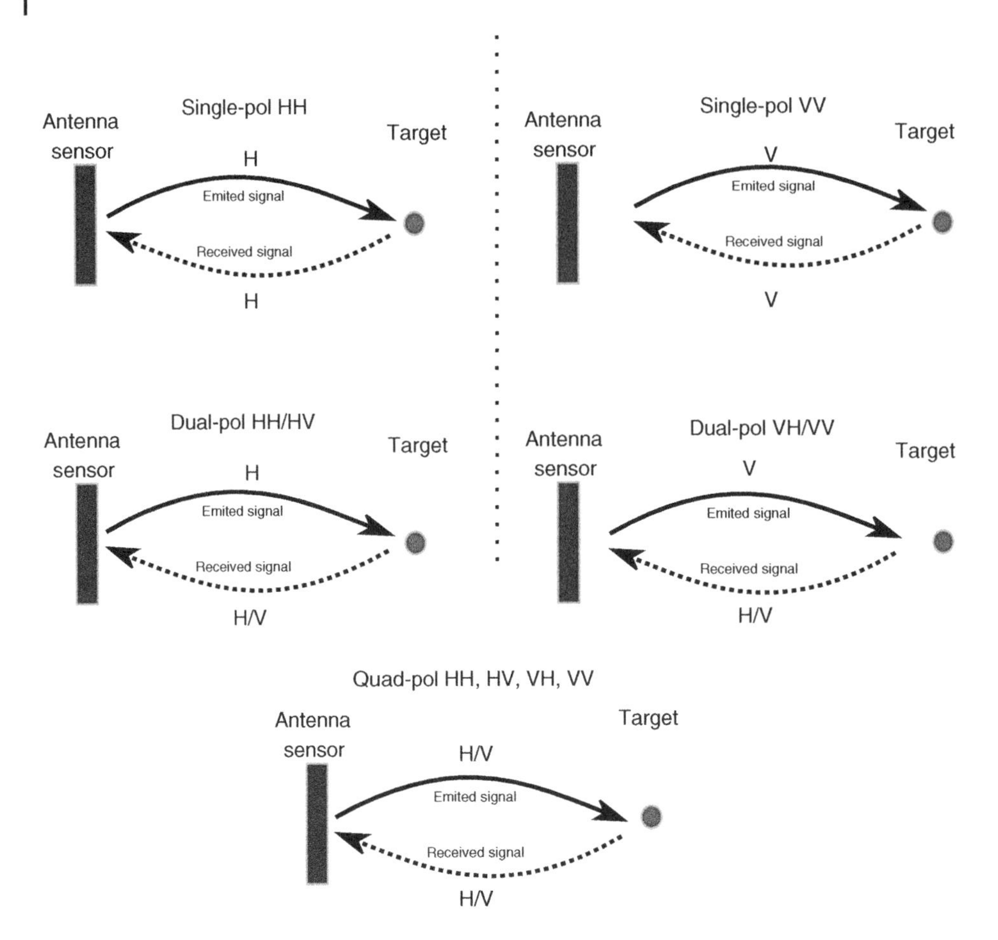

Figure 1.13 SAR common polarization modes.

ground objects that are, for instance, one meter apart as separate pixels within an image, the radar sensor needs to have the same spatial resolution: that is, one meter or less, which translates into the need of a large antenna (for the case of an orbiting satellite, such antenna would be of 15 km length!) if using a standard radar system. This is not the case, fortunately, when using SAR.

Figure 1.15 depicts a SAR system. First, note that the sensor is mounted on a moving platform (it can be a satellite, as the represented case, an airplane, a drone, as described by Koo and et al. (2012), or even a simple van, see Frey et al. (2013b)). For the case of the system shown in this figure, the radar system is mounted on a satellite, and it emits the electromagnetic signal to the ground (Earth surface), *illuminating* a region as it is moving along its orbit. The satellite is at an altitude (it could be around hundreds of kilometers or even much more), and it is moving following the radar trajectory. We show a *side-looking* observation system (other configurations are explained in Section 1.4). In this configuration, the radar is pointing perpendicular to the direction of flight. Two important concepts related to SAR imaging are also shown in this figure:

Slant range: is the side-looking direction of the antenna. It is perpendicular to the sensor's path (the along-track direction).

Azimuth: is the along-track direction, or also the perpendicular projection of the sensor line of flight on the ground.

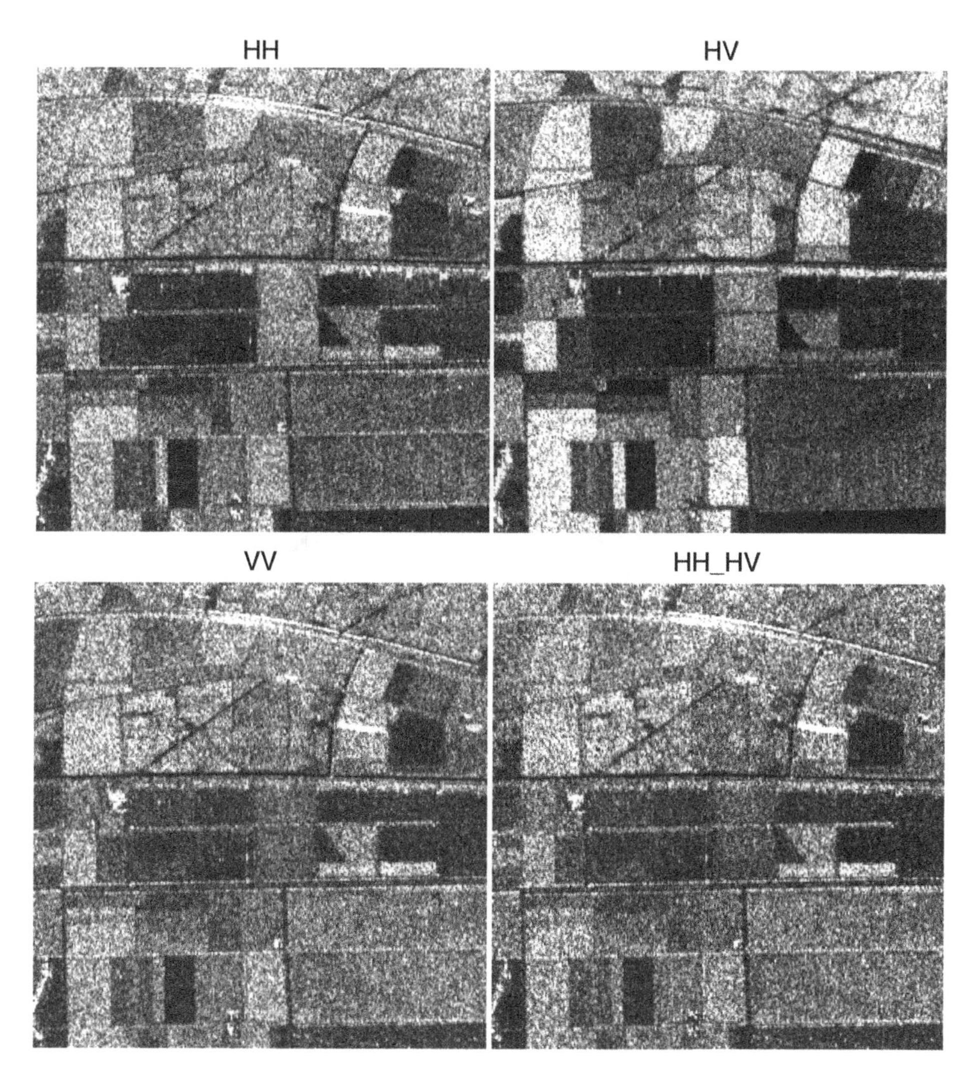

Figure 1.14 Example of polarization modes for an actual SAR image. The mode HH_VV indicates that the polarization used was quad-pol.

Therefore, the SAR antenna has its long axis in the flight direction (the azimuth direction) and the short axis in the range direction (slant range).

What defines a SAR system is the imaging technique employed. Figure 1.16 details the basic operation of a SAR system. As it can be seen, the forward motion of the antenna along the track direction is used to *synthesize* a much longer antenna. That is the reason for the system name SAR: Synthetic Aperture Radar[3]. At each antenna position, an electromagnetic pulse is emitted and the return echoes pass through the receiver and recorded for their post-processing. Figure 1.17 shows that the synthetic aperture (the sensor as it moves) collects many echoes from the same target. In this sequential way, the system operates as if it were using a larger antenna (the Synthetic Aperture Length). Figure 1.18 shows the synthetic aperture length. Note that the larger the antenna, the finer the detail the radar can

---

3 Aperture: from the Latin *apertus*, is translated as *open* or *opened*, and used in optics to name the opening through which light enters or exits an optical system. Note that microwaves are electromagnetic waves, so is light.

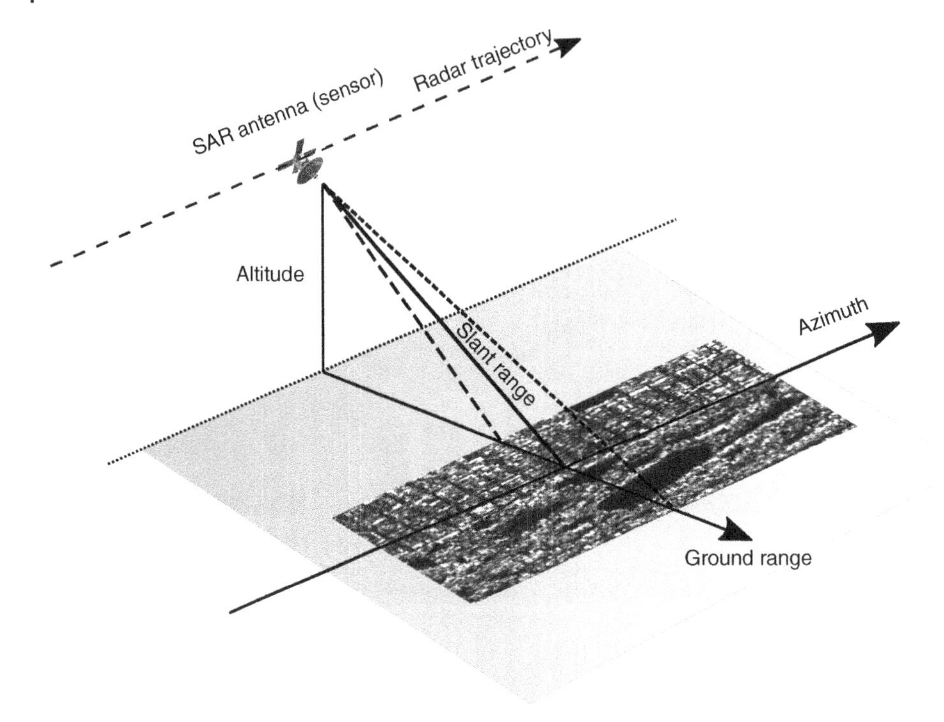

Figure 1.15 SAR acquisition geometry.

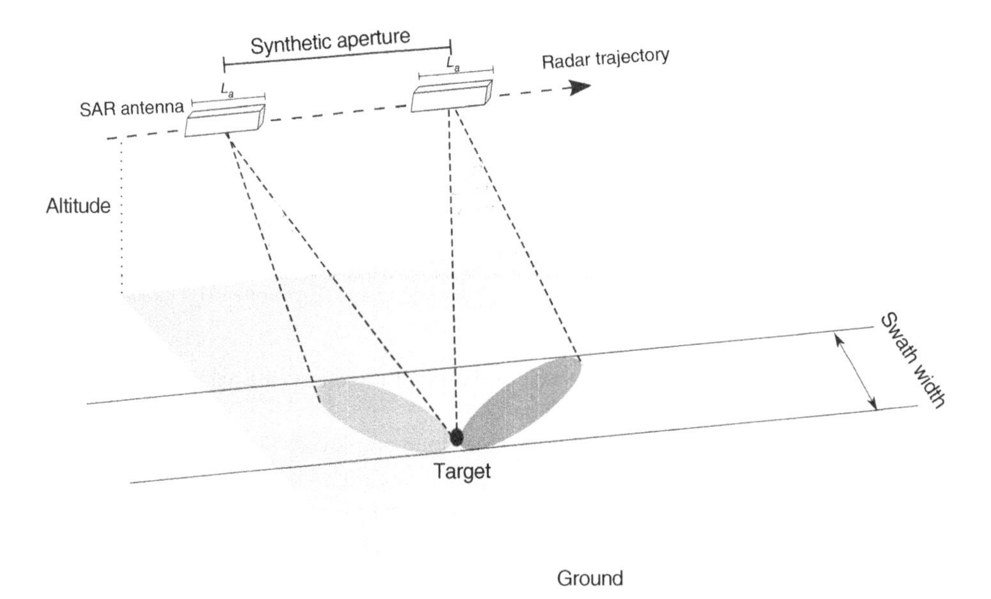

Figure 1.16 Synthetic aperture concept.

resolve. Therefore, by using a small antenna (which is a relevant issue when it is mounted on a moving platform), a high spatial resolution across large areas can be obtained.

From the stored echoes, the returned signal is converted to digital format and then processed to be later displayed. The processing is indeed complex, including many mathematical

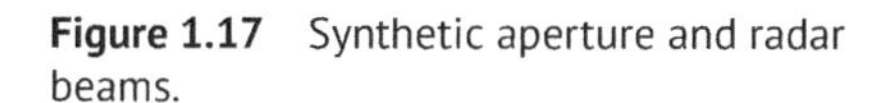

Figure 1.17 Synthetic aperture and radar beams.

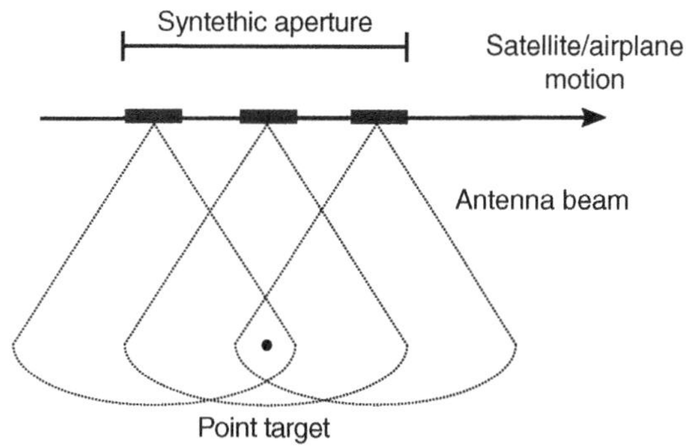

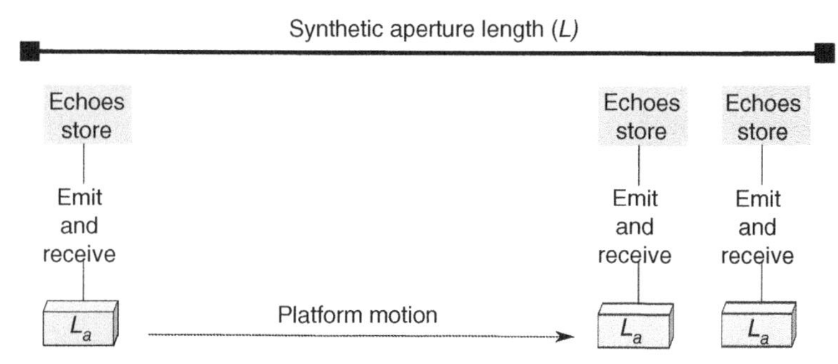

Figure 1.18 Synthetized antenna.

operations and specific signal processing methods that are mostly done in ground stations. The final result is a high-resolution image (see some actual SAR images in Figure 1.5) as it had been obtained by using a very long *physical*, antenna of length, for example, of 15 km or more.

It is interesting also to remark that, for a desired spatial resolution, the length of the antenna depends on the wavelength of the emitted signal. For instance, if using a C-band radar (which operates with a wavelength ≈ 5 cm), to obtain a spatial resolution of 10 m, a radar antenna longer than 4 km would be needed! Therefore, the benefits of the moving platform (satellite, aircraft or so) are clear.

The most common SAR bands and their main uses are collected in Table 1.1). More about spatial resolution is detailed in Section 1.3.

Table 1.1 Most used bands SAR.

Band	Frequency	Wavelength	Uses
X	8 GHz to 12 GHz	3.8 cm to 2.4 cm	High resolution (little penetration into vegetation): urban monitoring, ice and snow
C	4 GHz to 8 GHz	7.5 cm to 3.8 cm	SAR workhose (moderate penetration): change detection, global mapping, ocean observation
S	2 GHz to 4 GHz	15 cm to 7.5 cm	Agriculture monitoring and observation of high density vegetation areas
L	1 GHz to 2 GHz	30 cm to 15 cm	Medium resolution (high penetration): geophysical monitoring, vegetation and biomass mapping

1.3 Spatial Resolution

Spatial resolution is a key issue when dealing with acquired data. Lee and Pottier (2009) discuss in detail the following concepts.

Range resolution (slant range resolution): it is defined as the observable size of a ground pixel along the range direction (see Figure 1.19) and it is calculated as,

$$\delta_r = \frac{c\tau}{2}, \tag{1.2}$$

where c is the speed of light and τ is the duration of the transmitted radar pulse.

The projection on the flat ground plane of δ_r (see Figure 1.20) is known as the Ground Range resolution,

$$R_r = \frac{c\tau}{2\ \sin\theta}, \tag{1.3}$$

where θ is the look angle.

From the above expressions, note that:

- The range resolution is independent of the height of the platform (satellite, aircraft, etc.)
- The range resolution is infinite for an orthogonal direction of illumination (vertical look angle), and it improves as the look angle is increased.
- Finally, the range resolution improves as the bandwidth, BW, of the radar increases ($\tau = 1/\text{BW}$).

Azimuth resolution: it is defined as the observable size of a ground pixel along the azimuth direction (see Figure 1.19). The azimuth resolution, R_a, only depends on the length of the antenna,

$$R_a = \frac{L_a}{2}. \tag{1.4}$$

Therefore, the azimuth resolution is also independent of the height of the platform and improves as the antenna length is reduced.

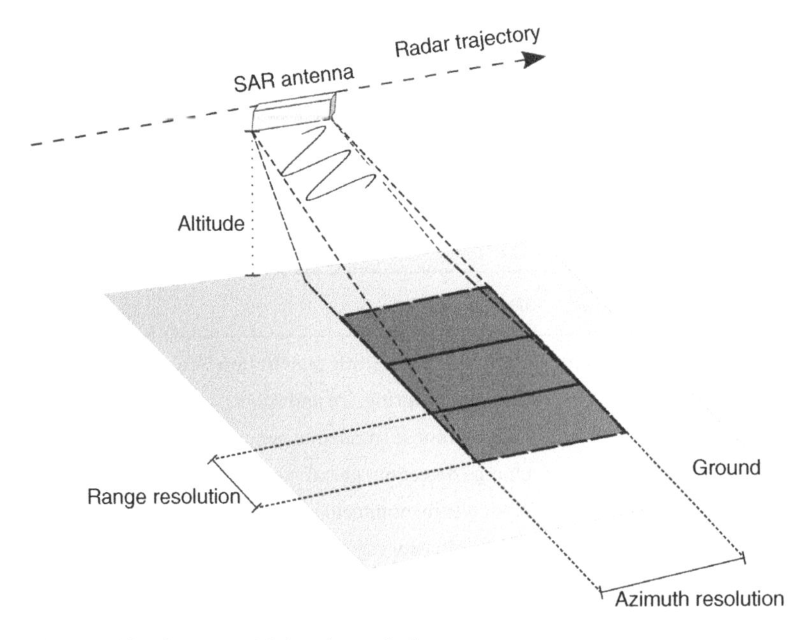

Figure 1.19 Range and Azimuth resolution.

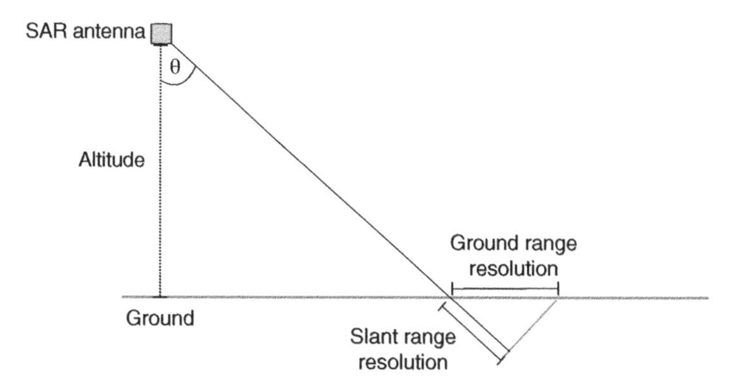

Figure 1.20 Slant range resolution and Ground range resolution.

1.4 SAR Imaging Techniques

In this section, a brief description of the most used SAR imaging techniques is provided. From the several available SAR acquisition modes, the most popular are:

Stripmap: (also known as *StripMap*): this corresponds to the conventional or original SAR mode (it was introduced in the early versions of SAR systems). As shown in Figure 1.21, the radar antenna is pointing at a fixed direction (fixed azimuth and observation angle) while moving along its flying path. Therefore, the ground swath is illuminated with a continuous sequence of pulses, the echoes are received and then, after processed, the image is formed. The azimuth resolution is notably increased because more echoes are received (multiple ones from the same target) than in a single scan. Strip mode is very useful to

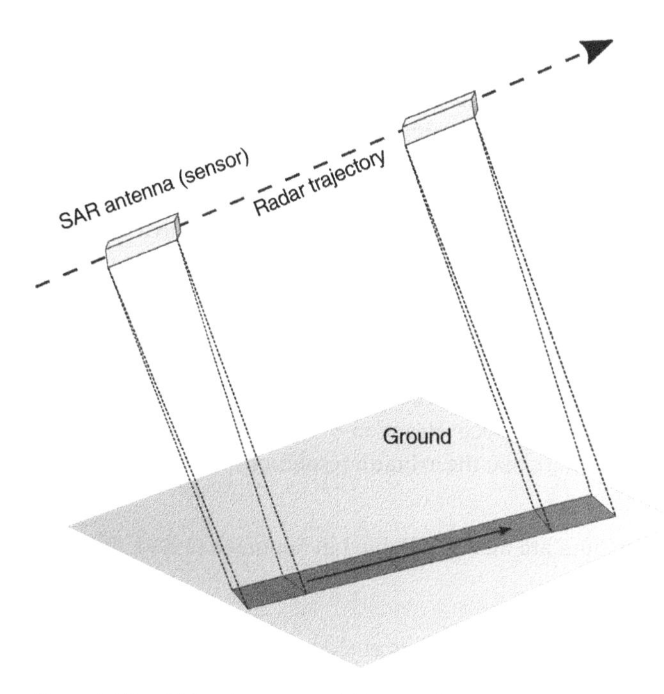

Figure 1.21 SAR stripmap acquisition mode.

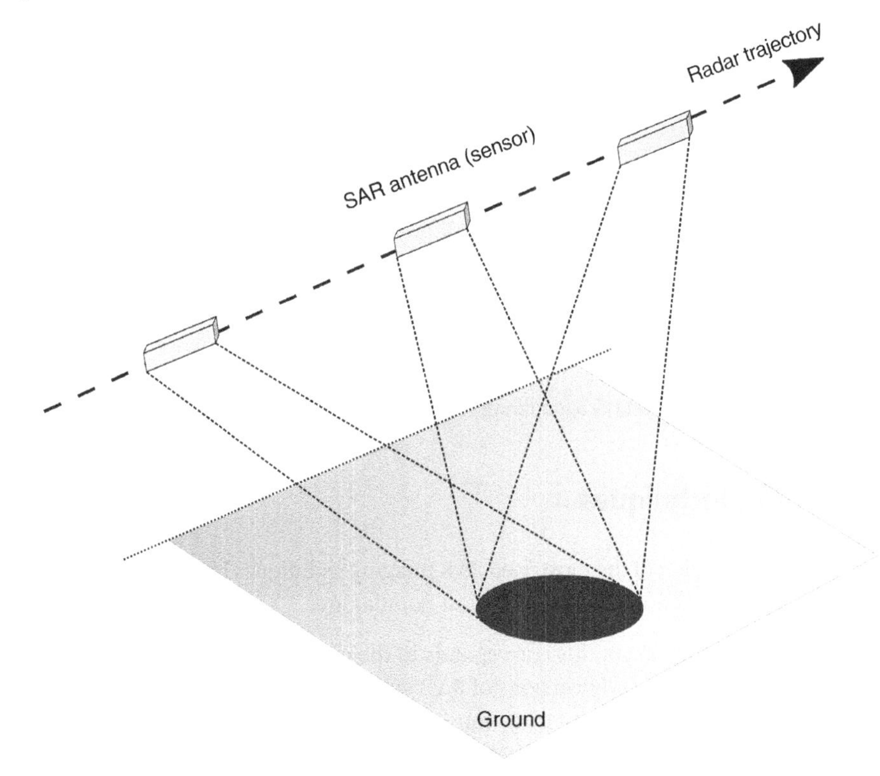

Figure 1.22 SAR spotlight acquisition mode.

obtain information of excellent resolution from large areas. Note that the radar antenna can be mounted on a satellite or on an aircraft.

Spotlight: (also known as *SpotLight*): this is the mode recommended for obtaining high-resolution images. As it can be seen in Figure 1.22, the antenna radar is pointing at a particular area and keeps illuminating just this area to get as many echoes as possible to increase both, the range and the azimuth resolutions significantly. Spotlight mode, as a difference from the stripmap mode, operates at the expense of spatial coverage. It is interesting to mention that the beam steering to irradiate on the specific area is either mechanically or electronically controlled to keep the radar radiation pattern constant on the target.

Scan: (also known as *ScanSAR*): in this mode, the radar antenna sweeps periodically into different orientations, thus, illuminating several subswaths areas (see Figure 1.23, covering a larger area than in the two previous modes explained above. Due to the synthetic aperture is shared between the subswaths areas, the reconstruction of the image requires mosaic operations. Also, it is important to remark that the azimuth resolution is much lower than for the stripmap and the Spot modes.

Stripmap and Spot imaging techniques are nicely explained in Soumekh (1999), through these examples (sic):

> "The stripmap SAR form of target area radiation is analogous to a scenario in which someone is trying to view *all* objects in a dark room with a flashlight. With the

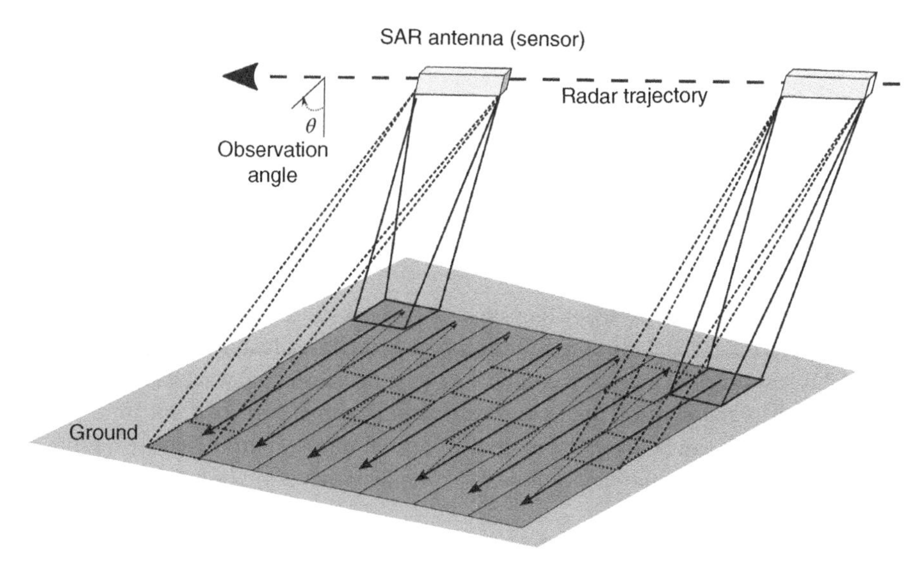

Figure 1.23 SAR scan acquisition mode.

flashlight in his right hand (the radar on the aircraft), that person would move his right arm to *scan* the room with the light of the flashlight. This *preliminary* phase of the search is to provide the individual with a general feel for what the room contains. Similarly, stripmap SAR systems provide imaging information on the general condition and contents of, for example, a terrain area.

Consider again the individual with the flashlight in a dark room. Once he locates an object of interest after scanning the beam of the flashlight throughout the room, he might go around that object with the flashlight to gather more information about it. This phase of search provides that individual with specific information about an object of interest, [...] a similar task is performed by spotlight SAR systems to obtain detailed information on targets within a relatively small terrain area."

Also in this text, rich information about SAR is found, but sound knowledge of signal processing is required.

1.5 The Return Signal: Backscatter and Speckle

Speckle is a term inherently related to SAR images, and it is largely studied in this book throughout statistical methods. This book's contents are all intended to deal with speckle. Before explaining what speckle is about, first, the radar return signal is briefly accounted.

1.5.1 Backscatter

An important concept is the radar cross section (RCS), which measures the ability to detect an object by the radar, that is to reflect (*backwards* to the radar), part of the energy transmitted by the radar toward the illuminated area. The RCS is also known as the electromagnetic signature of the object and is denoted by σ. It is influenced by many factors, such as,

- the size of the target and mainly, its relative size respect to the wavelength of the radar signal,
- the composition of the target,
- the incident angle of the radar emitted signal,
- the polarization of the radar emitted signal.

The amount of reflected energy by an object depends both on its physical properties and on the properties of the radar signal used. The reflected signal is known as the backscattered signal.

In Figure 1.24) the signal detected at the sensor (the backscattered signal) formed as a convolution of the reflected signals by each illuminated target is illustrated. Note that the reflected signal is just proportional to the value of σ, keeping its original waveform.

In Figure 1.25), it is represented the same information as in Figure 1.24 but in a three dimensional model. As it can be seen, the many scatters within the illuminated area, reflect the radar signal by modifying the amplitude of the incident waves proportionally to their backscatter coefficients σ.

To conclude this brief introduction to the backscatter coefficient, it is important to note that, as explained above, the amount of energy backscattered depends on the composition of the target. Without entering the physical properties (percentage of humidity, metallic

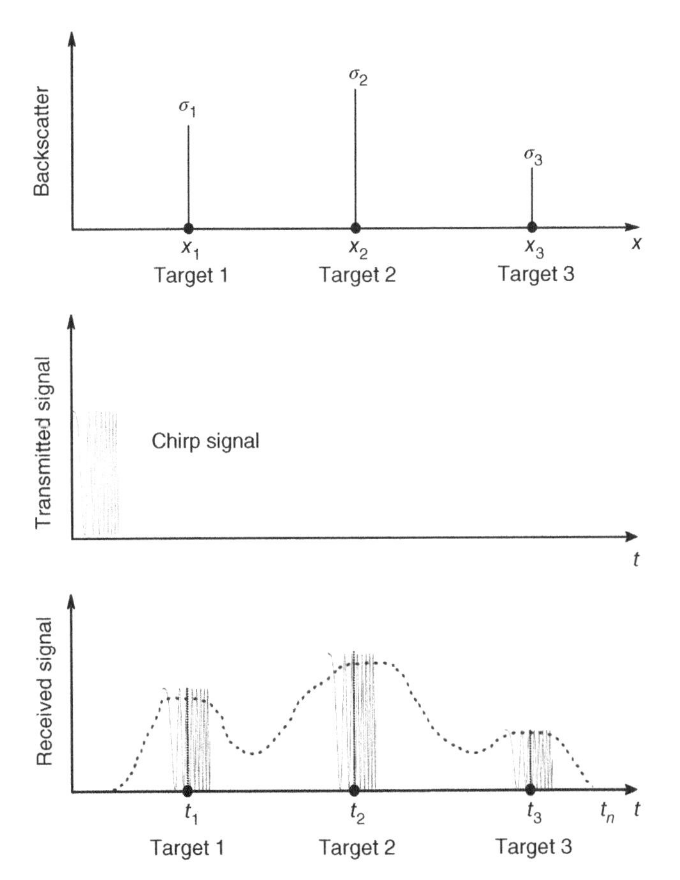

Figure 1.24 Unidimensional simplified model illustrating the detected signal (the echoed signal) from the convolution of the scattered signal by each target.

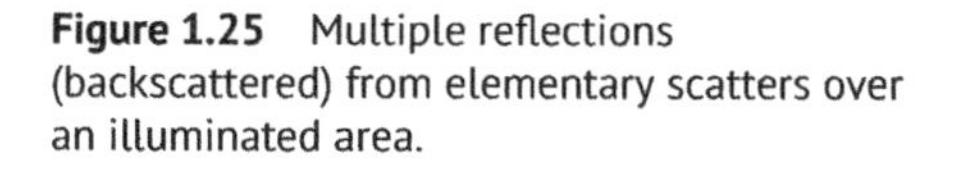

Figure 1.25 Multiple reflections (backscattered) from elementary scatters over an illuminated area.

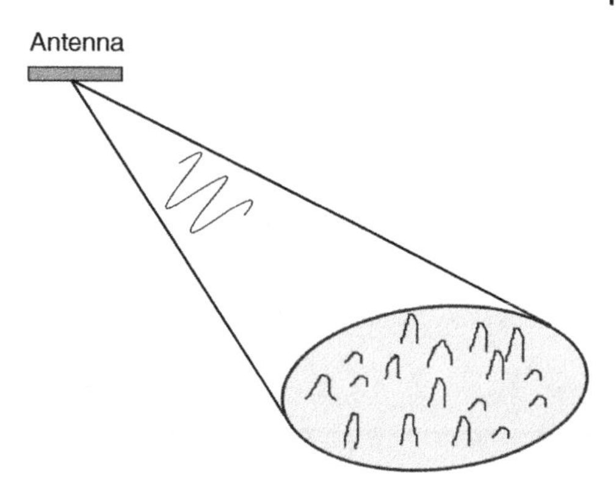

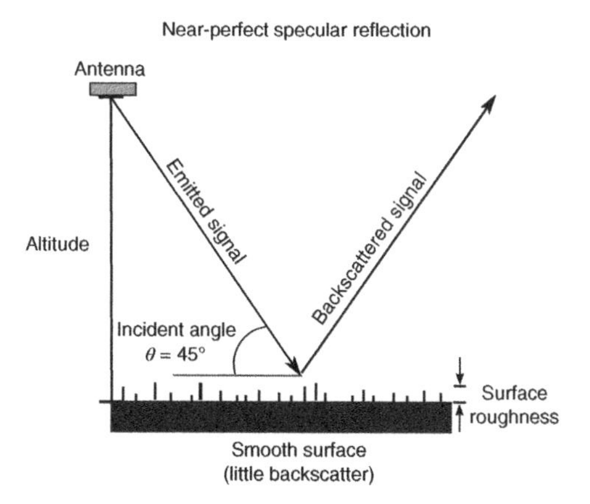

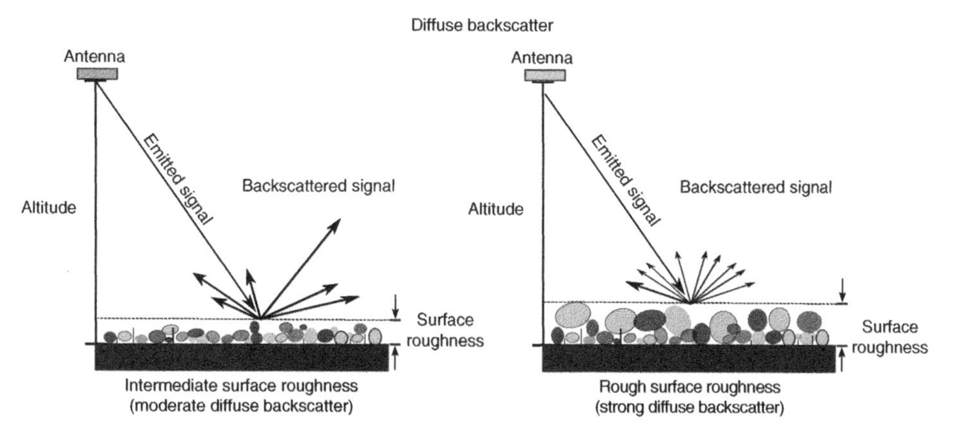

Figure 1.26 Backscatter and its relation to surface roughness.

components, etc.) and just focusing only on the size of the targets (when compared to the radar signal wavelength), three are the most important scattering mechanisms due to the surface roughness: the near-specular reflection, the near-diffuse, and the diffuse backscatter modes. These are shown in Figure 1.26).

In the case of near-specular reflection, no signal is received at the sensor, therefore, in the final image, it appears as a dark area (black pixels), as the black areas due to the calm ocean in

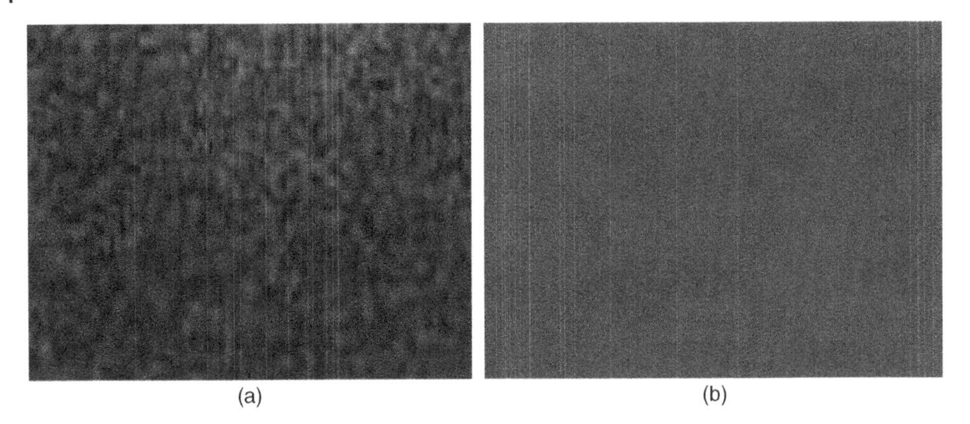

Figure 1.27 Zoom of the calm ocean from the San Francisco Bay SAR image (a), and the same area for the optical image (b).

Figure 1.6 (see a zoom of this image in Figure 1.27). The near and totally diffuse backscatter are due to reflections from objects of moderate and large size respectively and they will appear in the image brighter pixels (see, in this same image, the buildings and the forest area).

1.5.2 Speckle

Speckle statistical models are fully explained in Chapter 3. We address in this section the physical origin of speckle in SAR images.

An important characteristic of the radar signal in SAR systems is that they use a coherent signal[4]. A wave (signal, electromagnetic wave, light, etc.) is coherent when the phases of all waves at each point on a line normal to the direction of the beam propagation are identical. Usually, coherent light is monochromatic (single frequency) and a common practical source of such *light* is a laser or a radar signal. An example of coherent and incoherent signals (also known as non-coherent signal) is shown in Figure 1.28. From this image, it is clear

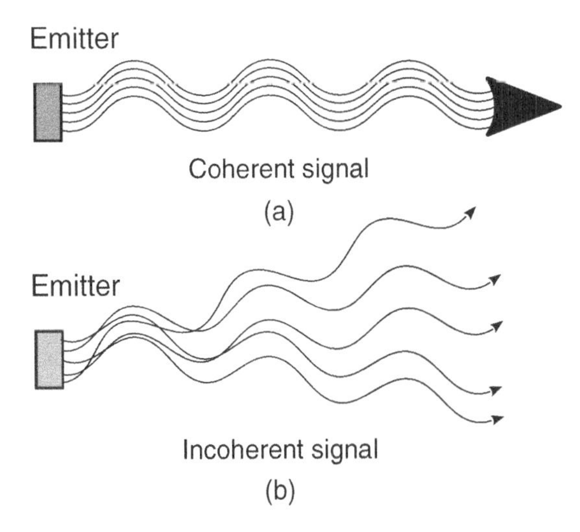

Figure 1.28 Coherent signal (a) and incoherent signal (b).

4 Not all radar systems use coherent signals (some systems use pseudo-coherent signals or also non-coherent signals).

that for the coherent signal, all phases are constant as the signal propagates. However, for the non-coherent signal, these phases are randomly distributed.

The benefits of using a coherent signal stem from the possibility of concentrating the emitted energy over the targets. Additionally, SAR data provides valuable information both in the amplitude and the phase of the received echoed signal. What defines a radar system as coherent or not comes from the type of transmitter used. The main advantage of a radar coherent system is that even very small phase shifts of the echo signals are detectable. Additionally, coherent radars also have a better signal-to-noise ratio than non-coherent systems. However, the use of coherent signals has an undesirable counterpart. After the the emitted signal interacts with the targets (see Figure 1.29), and from the fact that the used signal is coherent, backscattered signals may combine either in a destructive or a constructive way, or in a random combination of both definitive states. If this is extended to the real situation (as modeling in Figure 1.25) the consequence is that, from the multiple signal reflection due to the multitude of randomly located scatters, the received signal at the sensor is a sum of all those waved that, although coherent in frequency (all scattered signals have the frequency of the emitted radar signal), they are characterized by the absence of phase-coherence. Just as illustrated in Figure 1.29, for the case of destructive interference (waves are out of phase), a weak signal is received in the sensor. As for the constructive (waves are in phase), a strong signal is received. These mechanisms can be formulated mathematically through the sum of phasors (complex vectors) as it is explained in Chapter 3, and in more detail in the works by Yue et al. (2020, 2021b).

It is important to note that due to these mechanisms, many not-so-dark pixels will appear as black in the image. The same will occur to some not-so-bright pixels. This is the explanation of why the SAR images will look as corrupted with a salt-and-pepper noise. This characteristic granular pattern is what is known as speckle; it is easily visually recognizable. In a practical sense, a signal used in an imaging technique is considered to be non-coherent when no speckle effects are present in the final image and coherent when they are. The apparent

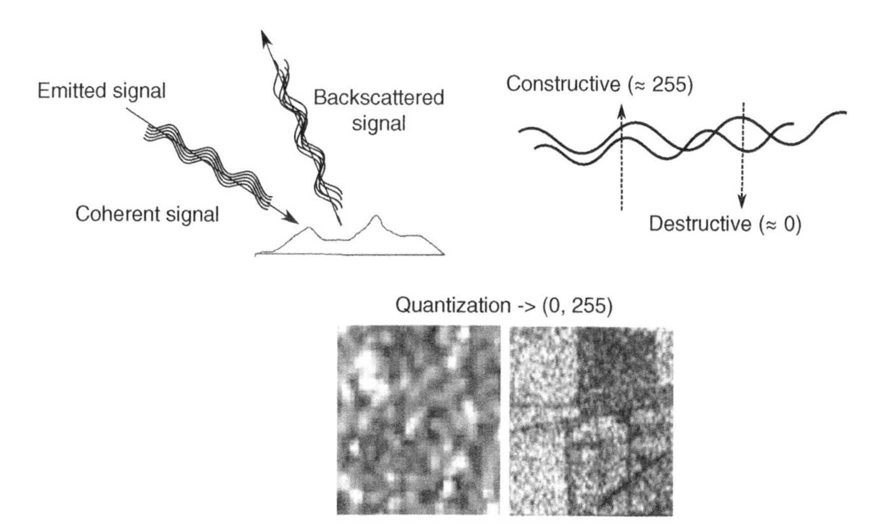

Figure 1.29 Constructive and destructive interference from using coherent signal. The constructive case (top right) will result in more white pixels (255), in a typical 8 bits quantization in the final image (bottom). The opposite occurs for the case of destructive case (more black pixels within the image). Both situations give the image a characteristic and easily visually recognizable granular pattern.

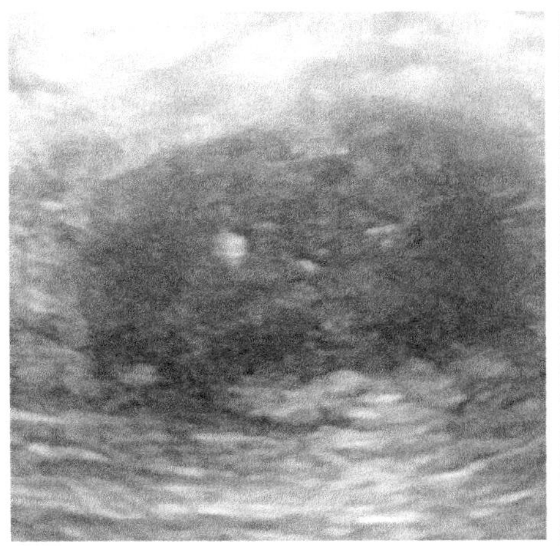
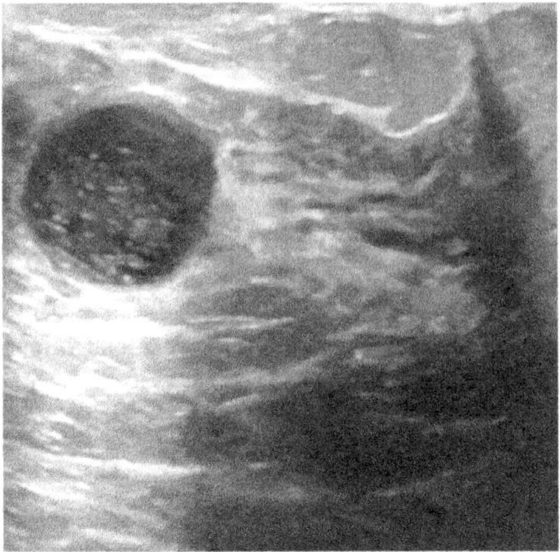

Figure 1.30 Actual breast ultrasound images. Source: From Breast Ultrasound Images Dataset, https://www.kaggle.com/aryashah2k/breast-ultrasound-images-dataset.

size of the scattering centers and that of the individual speckles observed in the image are related to the resolution limit of the acquisition system. Speckle is not exclusive of SAR images but it also appears in any imaging technique that uses coherent signals to form the image. Another well-known imaging technique largely affected by speckle is ultrasound-B (medical ultrasound images), laser, and sonar. An example of actual ultrasound images is shown in Figure 1.30.

From the above exposition, it is clear that speckle gives to final images a granular-like pattern, easily recognizable *visually* and it comes from illuminating the area under observation with a coherent signal. Speckle can hardly be considered noise as is the case of Gaussian noise in natural (optical) images, for it is not an added factor but a consequence of how the image is formed. If accepted that noise can be defined as a random or regular interfering effects that corrupt the quality of the data and the information it bears, speckle does not suit this definition. Regarding the particularities of the SAR imaging technique, speckle is a consequence of the resolution of the sensor, which is not sufficient to spatially separate (*resolve*) individual scatterers. Additionally, for the same acquisition conditions, an identical speckle pattern is obtained and, thus, it is not random.

In Goodman (1976), speckle is discussed from a physical point of view. This work is regarded as a seminal work, considered to be the first statistically modeling of speckle, which leads to the proposal of practical methods to reduce it. It can be said that the first speckle-reducing filters have been inspired by this work.

As explained above, each pixel in the final image is the result of the coherent superposition of all received signals from scatterers over the illuminated area. Indeed, from an image processing perspective, echoes within small parts of the illuminated area are jointly processed. Such small areas are termed *resolution cells*. The ensemble of the overall information from the resolution cells is what forms the final image. Besides, it can be assumed that the number and the distribution of the targets within each resolution cell are random (and unknown). A similar assumption holds for the physical composition of the targets and their sizes and textures. When one observes a SAR image, immediately will recognize its speckle content distributed within the whole image, showing a random alternation of bright and dark pixels due to the constructive and destructive patterns caused by the coherent illumination. Such

SAR image, being inherently a remote sensing image, typically will contain regions (usually large) related to homogeneous areas, such as agricultural soils, forest, roads, where the scatters within a resolution cell are almost uniform and, consequently, the speckle tends to be totally uncorrelated (randomly distributed). Such totally uncorrelated speckle areas (very common in SAR images) are known as fully developed speckle areas. However, this uncorrelated pattern is strongly modified by the geometric distortions due to the own nature of the SAR acquisition procedure (see Section 1.5.3); cf. Ref. (Yue et al., 2021a).

Although speckle contains information from the targets (indeed high-valuable information), speckle has a strong negative impact on the visual quality of SAR images, largely making it difficult their interpretation even for experts, not to mention the post-processing tasks (segmentation and classification), either manually or automatically (through machine learning techniques or deep learning approaches). Hence, speckle decreases the usability of SAR images since it makes it difficult to identify ground targets. Additionally, speckle content makes it difficult the estimation of the parameters that characterize the targets.

To mitigate the adverse effects provoked by speckle, many techniques have been researched over the past decades. Such efforts to reduce the speckle content within SAR images (and also in ultrasound medical images) is still ongoing. Basically, two are the techniques developed for despeckling (this is used term) SAR data:

The multilook approach consists of acquiring more images of the same illuminated area and to averaging all of them to reduce the speckle while preserving the amplitude level (Moreira, 1991). The approach is somehow similar to the HDR (High Dynamic Range) technique used in photography, where many photos are combined to improve the final result. In SAR, the images to be averaged correspond to images over the same illuminated area captured at different dates or taken from different look angles. A single-look SAR image indicates that one SAR image, whereas, for instance, a 4 Looks SAR image refers to a SAR image after averaging four (4) single-look SAR images. Although multilook technique notably improves the visual appearance of the image, it causes a loss of information caused by the averaging process. Indeed, multilook images seem blurred (the more blurred the higher the averages performed). That is, a single-look SAR image has a high speckle content (the maximum for that acquisition) but it also contains the maximum information (especially for edges and single targets). Examples of single-look SAR and 4-Look SAR images are shown in Figure 1.31. In summary, the multilook technique is easy to implement and it is very efficient in reducing speckle, but it also blurs the image in a similar way as a mean filter (boxcar filter) does. More on multilook technique, from a statistical perspective, is discussed in Chapter 3.

Filtering: reducing the amount of speckle by filtering the data (despeckling) is the other approach. The basic, but surprisingly very effective, median filter was first used and then replaced by more sophisticated filters. Along the past decades, many despeckling filters have been proposed, some of them, remarkably efficient (Lee et al., 2009). Essentially, a despeckling filter aims to reducing the speckle in large homogeneous areas while preserving image fine details (especially edges and bright scatterers, which are of particular importance in SAR). To that end, some methods (Lee et al., 2009, Kuan et al., 1985, Moschetti et al., 2006) explore pixels and close neighbors, and build local models from the data. Other methods (Cozzolino et al., 2014) explore larger areas (non-local filters). Yet other methods transform the data into a new domain through powerful imaging processing techniques, such as wavelet transform (Argenti and Alparone, 2002, Penna and Mascarenhas, 2019) or

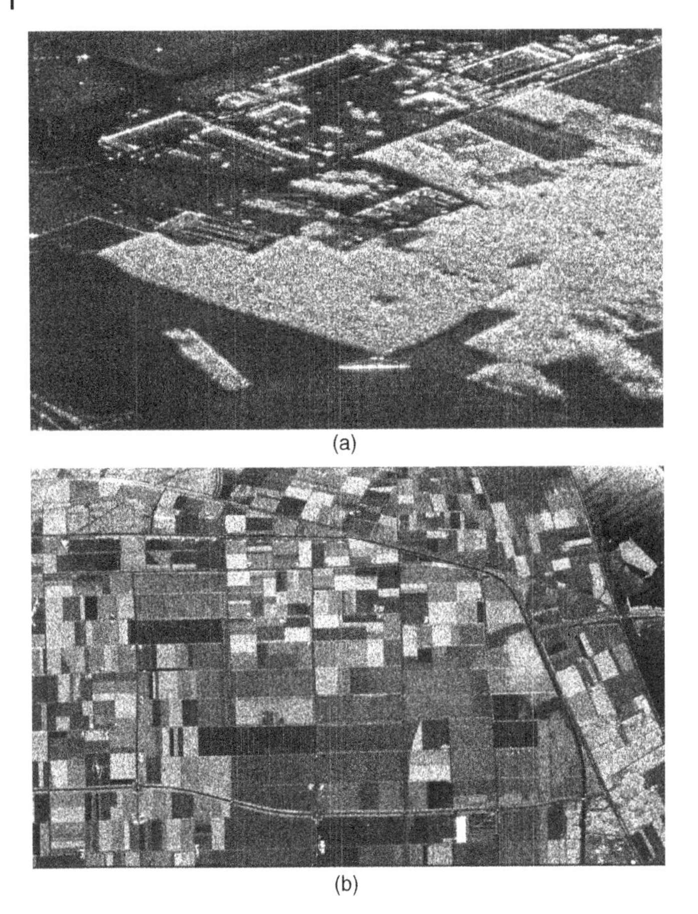

Figure 1.31 (a) Single Look SAR image (ESAR-HH, 1997 data) and (b) Flevoland (4-looks). In the latter, it can be identified the cultivated areas, some roads and even the small island. However, the interpretation of the single look image is a challenging task even for an expert.

total variation (Sun et al., 2021, Feng et al., 2014). The deep learning paradigm has irrupted recently into this challenging area with promising results (Ma et al., 2020). An example of despeckling a single-look SAR image by the simple median filter and the well-established Enhanced Lee filter is shown in Figure 1.32. Despeckling filters are studied in Chapter 5.

1.5.3 SAR Geometric Distortions

SAR systems, as often mentioned in this Chapter, are indeed complex system. Soumekh (1999), in the Preface, states

> "SAR is one of the most advanced engineering inventions of the twentieth century."

It is impossible in this Chapter to provide a whole explanation surrounding all technical aspects of SAR systems. Rather than that, our aim is reviewing those aspect which have a direct connection with speckle. SAR geometric distortions can be classified as:

- those caused by the side-looking nature of the SAR imaging system, which include effects as ground range nonlinearities, radar foreshortening, shadowing, and radar layover (explained below).

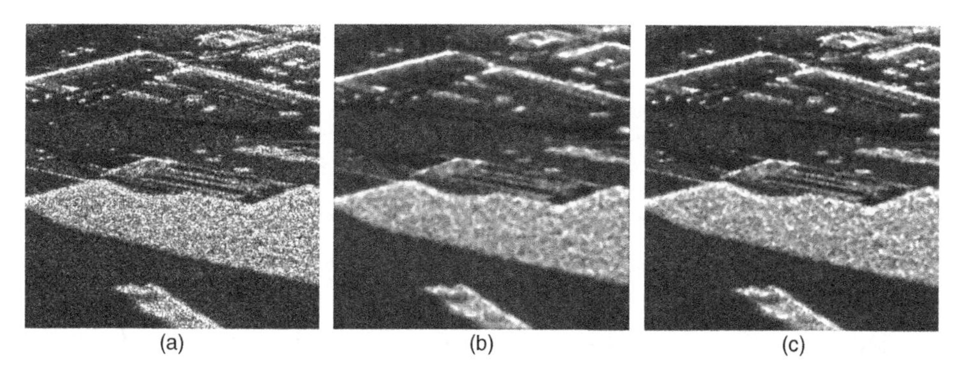

Figure 1.32 (a) Single Look SAR image (ESAR-HH, 1997 data) and (b) the filtered result by the median filter and (c) the result by the Enhanced Lee filter. It is clear that the latter performs better: image has more contrast (black areas appear blacker and bright areas, brighter) and fine details are better preserved.

- Distortion introduced during the data processing: it is due to approximations made to generate the final result (the image). They are mainly caused by errors in the estimation of the target phase or caused by wrong compensation for the Earth rotation during the acquisition time. The impact of these distortions depends on the precision of the algorithms used to obtain the final image.

In this section, only the first group of distortions are addressed. They are illustrated in Figure 1.33).

Details of the distortions caused by side-looking are:

Layover: the synthetic aperture (antenna) collects all the echoes from the targets within each resolution cell over the illuminated area. The ground coordinates are Azimuth and Range, being Azimuth the direction parallel to the flight movement of the antenna, and Range the orthogonal-to-movement direction. As the speed of the propagation of the emitted and received signals is finite (c, the speed of light), it is clear that targets along the range

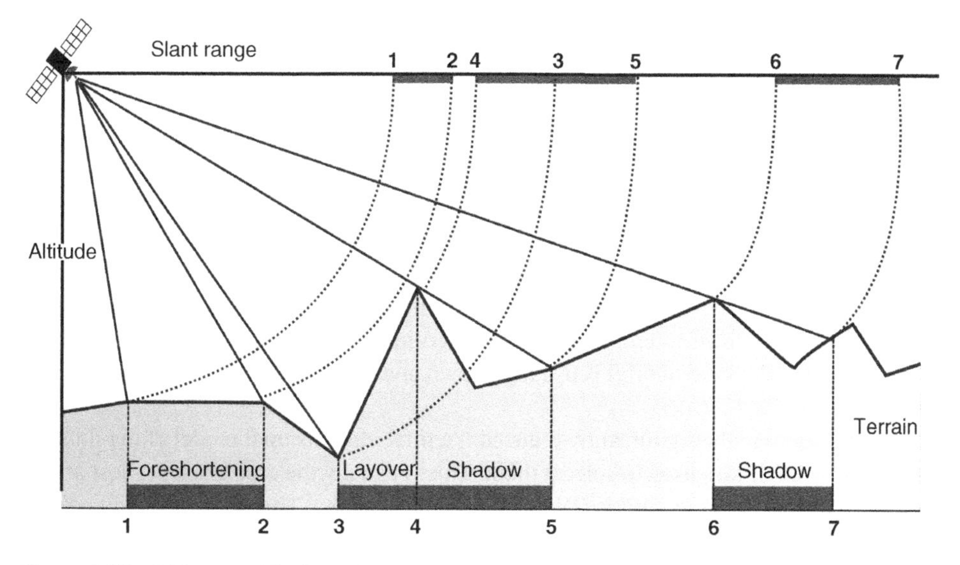

Figure 1.33 SAR geometric distortions.

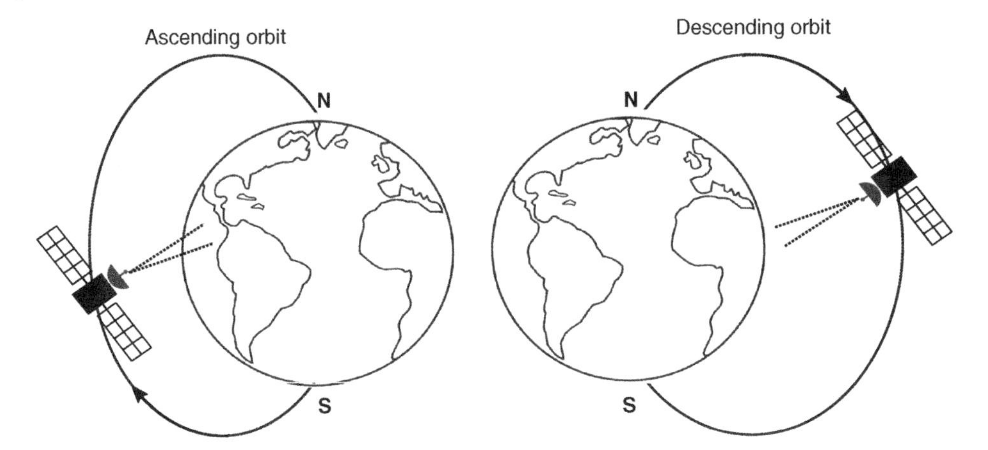

Figure 1.34 Ascending and descending polar orbits.

direction closer to the antenna will be captured first, so, in the final image, they will be represented also first. This is the desired situation for targets in the ground plane distributed with similar altitudes. However, for targets of different altitudes (see targets 3 and 4 in the ground plane in Figure 1.33), since the points located at a higher altitude are closer to the antenna than those at the base, they will be captured first and they will appear in the final image with a incorrect mapping (see, in the same figure, the mapping 3 → 4 and 4 → 3, when they should be mapped 3 → 3 and 4 → 4). This distortion is known as *layover*, and it is seen in SAR images that contain buildings or mountains. A SAR expert will visually recognize it easily.

Foreshortening: if the difference of altitudes among top and base points is not large enough, the mapping of the targets on the final image is correct but they will appear nearer (*compressed*) than they actually are in ground coordinates. This causes another SAR geometric distortion known as *foreshortening*. Such effect is represented in the figure (targets 1 and 2, which are correctly mapped, 1 → 1 and 2 → 2).

Shadowing: this is the last geometric distortion. See targets 4 and 5, and targets 6 and 7 in the figure. As its name suggests, the non-illuminated targets will not appear in the final image due to the absence of backscattering. For the case illustrated in this figure, the ground area from target 6 to target 7 will not be represented in the image and instead, a black area (black pixels) will be shown, which, obviously, does not agree with the actual scenario.

It is also important to note that two SAR images from the same illuminated area may show different kinds of geometric distortions mentioned above. That is, a shadow can be seen in one acquisition of an image and a different one in the other image due to the observation angle used in the acquisition. The same may happen for the foreshortening or/and the layover. This is because images can come from sensors with different observing angles. Note that some satellites operate in an ascending orbit and others in a descending one (see Figure 1.5.3). The information of the kind of orbit is usually known and it is useful when analyzing a SAR image.

These distortions can be significantly reduced if a three-dimensional model of the illuminated area is available and used to correct them. This is usually the case because most areas of interest have very accurate DEMs (Digital Elevation Models). Through image interpolation techniques, those corrections (terrain-corrected georeferencing) can be applied, at the

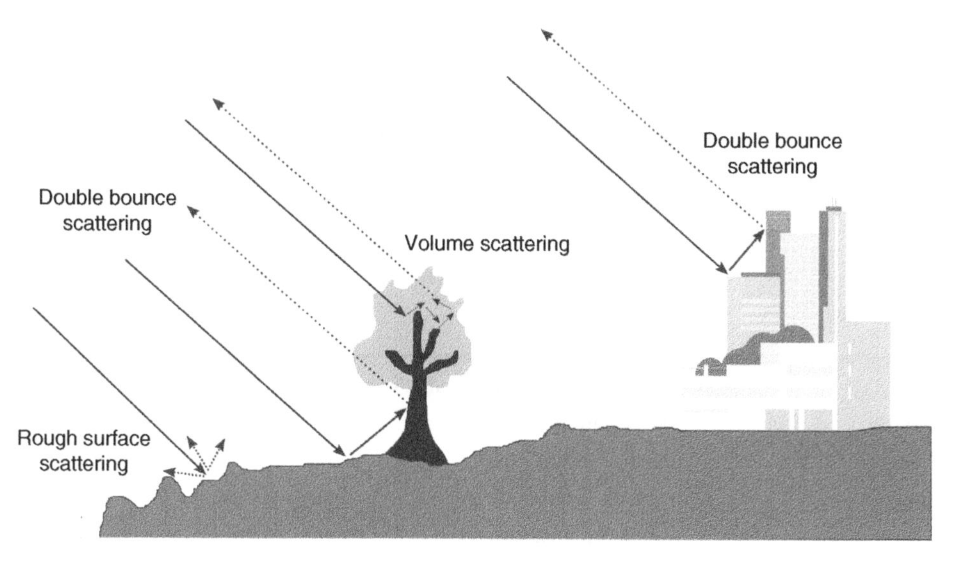

Figure 1.35 Scattering mechanisms.

cost of affecting the original spatial resolution (see, for instance, Frey et al., 2013a). Common software used for handling SAR images integrates these methods; cf. Section 1.7.

From these geometric interferences, a SAR image over a large illuminated area will commonly show the following types of areas:

- fully developed speckle regions, resembling a random pattern without visible structures. For the general case, the amplitude and the phase of the speckle field are statistically independent and the speckle phase are uniformly distributed over $(-\pi, \pi)$;
- in urban areas (with buildings), such regular pattern is less evident and it is replaced by a more complex pattern (shadowing and layover effects will manifest). Additionally, as illustrated in Figure 1.35, multiple reflections between ground and objects (trees, buildings, etc.) will be detected as stronger echoes in the sensor caused by the constructive interference from the reflected signals. Such double and multiple signal bounce scattering will be shown in the final image as very bright pixels, characteristic of SAR images. This scattering mechanism is, therefore, responsible for the illuminated side of objects which, in general, are mapped as strong bright pixels in the final image. It is also very common to identify shadows effects on the other side, along the range direction, where strong scatterers appear. Layover effects may cause the outstanding effect of mapping the buildings in non-natural disposition (the top of the building where the bottom should be or backward).

As mentioned, SAR images will show large variability, depending on the content of the illuminated area (see Figure 1.6, left). However, fully developed speckle areas and strong scatterers will play a decisive role when deciding the quality of a filtering operation performed on the image. This is discussed in the next chapters.

1.6 SAR Satellites

This section summarizes the SAR systems from the main Space Agencies. Special attention is given to spatial resolution and capabilities of those systems. Satellites equipped with SAR

Table 1.2 Overview of civilian SAR applications.

Area of application	Operation
Cartography	Topographic maps
	DEM (Digital Elevation Model)
Agriculture	Land cover monitoring
	Soil moisture measurement
Oceanography	Wind speed estimation
	Monitoring coastline changes
	Oil spill and maritime traffic surveillance
Hydrology	Flood monitoring
	Hydrological modeling
Geology	Landslide monitoring
	Geomorphological modeling
Forestry	Land use monitoring
	Detection of changes in vegetation
	Fire surveillance

Table 1.3 Some well-known SAR systems: a brief review.

Satellite	Country	Year	Band	θ (degrees)	Polarization	Range Resolution (m)	Azimuth Resolution (m)
SEASAT	EEUU	1978	L-Band	23°	HH	7.9	6
ERS-1/2	Europe	1991/95	C-Band	15° to 23°	VV	9.7	25
ENVISAT	Europe	2002	C-Band	15° to 45°	HH,HV,VH,VV	16.6	6

observation systems orbit the Earth in a sun-synchronous polar orbit, collecting data (amplitude and phase) day and night, independent of cloud coverage, with a revisiting period, of usually a few days. Table 1.2 presents an overview of some common civilian applications for SAR data.

Table 1.3 provides information about some well-known systems.

From this table, polarization modes have evolved from the single configuration (HH) to full polarization modes (HH, HV, VH, VV). The spatial resolution (both, in range and in azimuth) has also been notably improved.

Although this section focuses on main SAR systems, which are detailed above, for the sake of completeness, other relevant agencies (and SAR missions) are:

- Indian Space Research Organization (ISRO): RISAT-1
  - Year: 2012.
  - Polarization: dual.
  - Resolution: 3 to 50 meters, and one meter in Spotlight mode.
- Argentina's space agency (CONAE, Comisión Nacional de Actividades Espaciales): SAOCOM,
  - Year: 2018 (SAOCOM 1A), 2020 (SAOCOM 1B).

  - Polarization: HH, HV, VH, VV.
  - Resolution: 7 to 100 meters (SAOCOM 1A),.
- Korea Areospace Research Institute (KARI): KOMPSat-5,
  - Year: 2013.
  - Polarization: HH, HV, VH, VV.
  - Resolution: 20 meters (Stripmap mode) and one meter at high resolution mode.

We now provide details about the main SAR systems. We first present the European and American missions, followed by the Japanesse and Chinesse missions. It is recommended to visit the ESA (European Space Agency) website, (https://www.esa.int/) and, particularly, the Sentinel website (https://sentinel.esa.int/web/sentinel/home). For the COSMOS-Sky Med, it is recommended to visit (https://directory.eoportal.org/web/eoportal/satellite-missions/c-missions/cosmo-skymed) and for the RADARSAT, https://www.asc-csa.gc.ca/eng/satellites/radarsat/default.asp. For the TerraSAR-X, it is recommended to visit the website https://www.dlr.de/content/en/missions/terrasar-x.html, and for NASA's airborne and satellite-based systems, the website https://earthdata.nasa.gov/learn/articles/getting-started-with-sar. See the website https://directory.eoportal.org/web/eoportal/satellite-missions/a/alos-2 for more information about the Japanese mission, and https://eoportal.org/web/eoportal/satellite-missions/g/gaofen-3 for the Chinese mission. All the information summarized in this section has been extracted from those websites.

1.6.1 European Mission: Sentinel-1

The Sentinel missions include radar (SAR) and super-spectral imaging for land, ocean, and atmospheric observation and monitoring. Sentinel is part of the European Union's Copernicus program. Copernicus includes many missions carrying different sensors (Sentinel-1, Sentinel-2, Sentinel-3, Sentinel-4, Sentinel-5, and Sentinel-6), and with many near-future missions (Sentinel-7, Sentinel-8, Sentinel-9, Sentinel-10, Sentinel-11, and Sentinel-12) to be included in the second generation of Copernicus program. Sentinel-1 uses a SAR system and it is the one summarized in this section. Sentinel-1 mission consists of two twin polar-orbital satellites (*a constellation*) sharing the same orbital plane: the Sentinel-1A satellite (launched in 2014), and the Sentinel-1B satellite (launched in 2016), operating day and night. Both satellites carry a C-band SAR, which provides high-resolution imagery with high cloud penetrating capabilities. Sentinel-1 mission has a short revisit period, making it reliable for repeated wide-area monitoring. This is of particular interest for detecting and tracking oil spills and to mapping sea ice, as well as for detecting landslides and changes in the land uses. Through the Copernicus Hub (see Section 1.7), Sentinel-1 data are freely accessible. For each observation, precise measurements of spacecraft position and attitude are available. In particular, no distinction is made between public, commercial, and scientific uses, or between European or non-European users. Each Sentinel-1 satellite is expected to transmit Earth observation data for at least 7 years and have fuel on-board for 12 years.

Table 1.4 gathers a selection of the Sentinel-1 data of interest.

The SAR instrument onboard in the Sentinel satellites operate in one of the four exclusive modes:

Stripmap: the sensor provides continuous coverage with a ground resolution of 5 m × 5 m at a swath width of 80 km while the antenna beam is pointing to a fixed azimuth and elevation angle. This mode will only be operated on request for extraordinary situations.

Interferometric wide swath: this is a new type of Scan mode, called Terrain Observation with Progressive Scan (TOPSAR, with the purpose to reduce the drawbacks of the standard

Table 1.4 Some Sentinel-1 parameters.

Sentinel-1 parameter	Value
Polarization modes	Single (HH or VV)
	Dual (HH + HV or VV + VH)
Antenna length	12.3 m
Antenna width	0.82 m
Azimuth beam width	0.23°
Elevation beam width	3.43°
Elevation beam steering range	−0.9° to 0.9°
Maximum range bandwidth	100MHz
Apogee altitude	683 km
Repeat interval	12 days (175 orbits/cycle)

Scan mode. The basic principle of TOPSAR is the shrinking of the azimuth antenna pattern (along-track direction) as seen by a spot target on the ground. This is obtained by steering the antenna in the opposite direction as for Spotlight support. This acquisition mode is Sentinel-1's primary operational mode over land.

Extra wide swath: an extension of the interferometric wide swath mode to cover very wide areas (400 km) at a medium ground resolution of 20 m × 40 m.

Wave: in this mode, a set of stripmap scenes (known as *vignetes*), at regular spatial intervals of 100 km along-track are acquired. This mode is particularly well-suited for ocean applications, where data can be taken over large areas at low resolution.

1.6.2 European Mission: COSMO-SkyMed Systems

COSMO-SkyMed (Constellation of Small Satellites for Mediterranean basin Observation) is a 4-satellite constellation, operated by ASI (Agenzia Spaziale Italiana). The COSMO-SkyMed program, launched in 2007, represents, at present, the largest Italian investment in space systems for Earth Observation. Each satellite is equipped with an X-band SAR instrument, and the main goal of the mission is to provide global Earth observation for both, the military community and the civilian society (commercial and for institutions). The four satellites are in sun-synchronous polar orbits with a nominal altitude of 619 km and an orbital period of 97.2 min. The operating life of each satellite is estimated to be of five years, and the revisit period is of 16 days. The radar antenna is 1.4 m wide and 5.7 m long, operating at a center frequency of 9.6 GHz with a maximum radar bandwidth of 400 MHz. The polarization modes available are the same as for the Sentinel-1 mission: single and dual. Table 1.5 presents a selection of the COSMO-SkyMed data characteristics.

1.6.3 European Mission: TerraSAR-X

TerraSAR-X has been a remarkable European mission for observing the Earth promoted by the German Aerospace Center (DLR) and EADS (European Aeronautic Defence and Space Company) Astrium. The SAR satellite was launched in 2007 and became fully operational in 2008. It orbiting the Earth in a polar orbit at 514 km altitude. Then, in 2010, a twin satellite complemented the initial system, that became to be known as the TanDEM-X system,

Table 1.5 Cosmo-SyMed acquisition modes.

Mode	Swath km × km	Resolution (m) Range × Azimuth	Polarization
Stripmap (Himage)	40 × 40	3 × 3	Single
			HH or VV or HV or VH
Stripmap (Ping-Pong)	30 × 30	15 × 15	Alternating
			HH/VV or HH/HV or VV/VH
Spotlight-2	10 × 10	1 × 1	Single
			HH or VV
Scan (huge area)	200 × 200	100 × 100	Single
			HH or HV or VH or VV
Scan (wide area)	100 × 100	30 × 30	Single
			HH or HV or VH or VV

which acquires data for the worldwide and homogeneous DEM (Digital Elevation Model) high-valued data for mapping the Earth's surface. TerraSAR-X (also referred as TSX mission or TerraSAR-X1) is intended for both, scientific and commercial applications, operating at multi-mode and high-resolution X-band. Terra-SAR-X data have wide applications: hydrology, climatology, disaster monitoring, cartography (DEM generation of 2D and 3D maps) when it operates in interferometry and stereometry conditions, among others. Due to its short revisiting time (2.5 days) and access to any point on Eath, TSX mission is of enormous value for change detection of land use or even for monitoring large-scale construction projects. Some characteristics of the TerraSAR-X system are collected in Table 1.6.

Table 1.7, displays some parameters regarding the imaging modes of the TerraSAR-X system.

A collection of TerraSAR-X data is freely available for users at https://tpm-ds.eo.esa.int/oads/access/collection/TerraSAR-X.

1.6.4 Canadian and NASA Missions

First, the RADARSAT Mission is summarized, and then, some NASA missions are also briefly addressed. The RADARSAT Constellation Mission (RCM), launched in 2019, is Canada's new generation of observation satellites for scanning the Earth day or night in any weather

Table 1.6 TerraSAR-X system characteristics.

TerraSAR-X Parameter	Value
SAR antenna dimensions (length × height)	4.8 m × 0.80 m
Centre frequency	9.65 GHz ($\lambda = 3.1$ cm)
Resolution (maximum)	1 m
Polarization	HH, VV, HV, VH
	single or dual
Revisiting time	11 days

Table 1.7 TerraSAR-X acquisition modes.

Beam Modes	Scene size (width × length)	Maximal spatial resolution (m)
Stripmap	30 km × 50 km	3
Spotlight	10 km × 5 km	1
Scan	100 km × 150 km	16

conditions by using a constellation of three identical SAR satellites operating in C-Band. This constellation allows for a daily revisit of Canada's territory. It also allows daily access to $\approx 90\ \%$ of the planet and up to four times a day for the Artic.

RCM is the natural evolution of the previous RADARSAT program based on the RADARSAT-2 satellite (launched in 2007, yet active), but fully optimized (including a strong reduction of its size: half the size of its predecessors).

RCM data is used in many fields: maritime surveillance, agriculture, climate change monitoring, ecosystem monitoring, land use evolution, and even human impact on the Earth's surface.

Although RCM data are mostly restricted, certain image products are freely and openly available for worldwide users, under certain conditions.

However, for RADARSAR-2 mission, data is freely available after registration (see the website https://earth.esa.int/eogateway/missions/radarsat).

A full comparison of characteristics of RADARSAT-2 and the RCM is available on the website (https://www.asc-csa.gc.ca/eng/satellites/radarsat/technical-features/radarsat-comparison.asp), and Table 1.8, gathers some of those parameters.

Table 1.9 presents some parameters regarding the resolution capabilities of the RCM system. Similar parameters for the RADARSAT-2 are gathered in Table 1.10.

Apart from the Canadian mission, NASA's airborne and satellite systems provide also high-valuable SAR data for users (many of them freely). In Section 1.8, we present a brief introduction to some useful tools to access NASA's data.

It is also worth mentioning the NASA-ISRO SAR (NISAR) Mission (https://nisar.jpl.nasa.gov/mission/quick-facts/, to be launched in 2022 from India's Satish Dhawan Space Center in Sriharikota, India, into a near-polar orbit, which the purpose of monitoring climate

Table 1.8 RADARSAT-2-RCM system characteristics.

	Item RADARSAT-2	RADARSAT constellation
Altitude	798 km	586–615 km
High resolution (spotlight mode)	1 m × 3 m	1 m × 3 m
SAR antenna dimensions (length × height)	15 m × 1.5 m	6.75 m × 1.38 m
Centre frequency	5.405 GHz ($\lambda = 5.7$ cm)	5.405 GHz ($\lambda = 5.7$ cm)
Bandwidth	100 MHz	100 MHz
Active antenna	C-Band	C-band
Polarization	HH, VV, HV, VH	HH, VV, HV, VH Compact

Table 1.9 RCM acquisition modes.

Beam Modes	Nominal swath width (km)	Approximate resolution (m)
Low Resolution (100 m.)	500	100 × 100
Medium Resolution (50 m.)	350	50 × 50
Medium Resolution (30 m.)	125	30 × 30
Medium Resolution (16 m.)	30	16 × 16
High Resolution (5 m.)	30	5 × 5
Very High Resolution (3 m.)	20	3 × 3
Ship Detection	350	variable

Table 1.10 RADARSAT-2 acquisition modes.

Beam Modes	Nominal swath width (Km)	Maximal spatial resolution (m)
Standard	100	25
Wide	150	25
ScanSAR Narrow	300	50
ScanSAR Wide	500	100
Ocean Surveillance	530	Variable

change. This mission will scan the Earth's surface (land and ice-covered areas) with a visiting time of 12 days with a maximum resolution of 3 m. NISAR will become the first SAR satellite to use two microwave bandwidth regions: L-band ($\lambda = 24$ cm) and S-band ($\lambda = 9$ cm). By using such short wavelengths, this sensor will be able to observe changes in the Earth's surface of less than a centimeter in size.

1.6.5 Japanesse Mission

Japanese SAR systems include the L-band satellites under the general name ALOS (Advanced Land Observing Satellite): ALOS/AVINIR-2, PALSAR) and ALOS-2 (ALOS-2/ScanSAR). Data from these satellites, since 2019, are freely available. The main objective is to provide data continually for regional observation, disaster monitoring, environmental observation, and cartography. The ALOS/PALSAR (Advanced Land Observing Satellite Phased Array L-band Synthetic Aperture Radar) is an enhanced version of the JERS-1 (Japanese Earth Resources Satellite 1). It was launched in 2006, and it operates in a sun-synchronous orbit at an altitude of 691 km, with a 46-day recurrence cycle. The PALSAR sensor operates in a wide range of off-nadir angles and resolutions in a single-, dual-, and quad-pol mode.

The ALOS-2 SAR was launched in 2014 in a Sun-synchronous orbit, at an altitude of 628 km, with a revisiting time of 14 days. Some specifications for this satellite are collected in Table 1.11.

Table 1.12 presents some parameters regarding the resolution capabilities of the ALOS-2 system.

Table 1.11 ALOS-2 System Characteristics.

ALOS-2 Parameter	Value
SAR antenna dimensions (length × height)	9.9 m × 2.9 m
Centre frequency	9.65 GHz (λ = 3.1 cm)
Band (wavelength)	L-band (λ = 22.9 cm)
Resolution (maximum)	1 m (range) × 3 m (azimuth)
Polarization modes	Single / dual / full / compact

Table 1.12 ALOS-2 acquisition modes.

Beam Modes	Swath	Spatial resolution (m)
Stripmap	Ultra-fine: 50 km	3
	High-sensitive: 50 km (FP: 30 km)	6
	Fine: 70 km (FP: 30 km)	10
Spotlight	25 km × 25 km	3 (range) × 1 (azimuth)
ScanSAR	350 km	100

1.6.6 Chinese Mission

The Chinese Academy of Space Technology developed the Gaofen systems (Gaofen 1, Gaofen 2, up to Gaofen 14), and many powerful Gaofen satellites are in space since the launch of the first one, Gaofen 1, in 2013. The latter was a high-resolution optical observation satellite, and, in general, the main goal of the Gaofen mainframe relies on performing observations for disaster prevention and relief, geographical mapping, climate change monitoring, environment, and resource surveying, and also for precision agriculture support. However, Gaofen-3 is a SAR civilian satellite, that operates in the C-band, aiming to provide high-resolution images (land and ocean monitoring, disaster reduction, water conservancy, and meteorology) and disaster monitoring. It is in a Sun-synchronous orbit (like the RADARSAT-2 and Sentinel-1 systems), at an altitude of 755 km. Table 1.13 collects some of its parameters.

Table 1.13 Gaofen system characteristics.

Item	Value
Band	C-band
SAR antenna dimensions (length × height)	15 m × 1.5 m
Polarization	Single/Dual/Full
Acquisition modes	12
Swath width	10 km to 650 km
Spatial resolution	1 m to 500 m

Table 1.14 The 12 acquisition modes of Gaofen-3.

Acquisition Mode	Observation angle (°)	Resolution (m)	Swath width (km)	Polarization
Spotlight	[20, 50]	1	10×10	single
Ultra-fine Stripmap	[20, 50]	3	30	single
Fine stripmap	[19, 50]	5	50	dual
Wide fine stripmap	[19, 50]	10	100	dual
Standard stripmap	[17, 50]	25	130	dual
Narrow ScanSAR	[17, 50]	50	300	dual
Wide ScanSAT	[17, 50]	100	500	dual
Global observation	[17, 53]	500	650	dual
Quad-pol stripmap	[20, 41]	8	30	quad
Wide quad-pol stripmap	[20, 38]	25	40	quad
Wave	[20, 41]	10	5×5	quad
Expanded incidence angle	[10, 20]	25	130	dual
	[50, 60]	25	80	dual

Table 1.14 presents some parameters regarding for the twelve (12) observing modes of Gaofen-3, provided by the C-SAR (Complementary SAR) sensor.

1.7 Copernicus Open Access Hub

Information summarized in this section is from the website https://scihub.copernicus.eu/. To have a global understanding of the Copernicus program, it is recommended to visit the website https://www.copernicus.eu/en. According to those websites,

> Copernicus is the European Union's Earth observation program, looking at our planet and its environment to benefit all European citizens. It offers information services that draw from satellite Earth Observation and in-situ (non-space) data

It takes its name from the great well-known European scientist and astronomer Nicolaus Copernicus.

Copernicus program provides a vast amount of global data from different systems: satellites, ground-based, airborne and, seaborne, free and openly accessible to users.

How to access satellite data? Fortunately, this is an easy task using the Copernicus Open Access Hub, which provides complete, free and open access to data from Sentinel-1 (SAR) and other products (Sentinel-2, Sentinel-3, and Sentinel-5P).

For the case of interest of this book (SAR data), a brief description of Sentinel-1 data available from the Copernicus Open Access Hub is detailed below. Readers are strongly encouraged to register and experiment with the Copernicus' Hub to discover its enormous possibilities. Through an interface, it is easy to map any zone under satellite observation and to request for products (data), download them, and start using them. The data are free of charge to all data users (commercial users, scientific and general public) under the Sentinel Standard Archive Format for Europe (SAFE) format which is recognized by ESA tools (also free) like SNAP (see the end of this section). Data are available in single-polarization (VV or HH) for the Wave mode and in dual-polarization (VV+VH or HH+HV) or single-polarization for Stripmap, Extra Wide Swath mode, and Interferometric Wide Swath mode. The data are organized in levels:

Level-0: this can be considered the most complex data, not indeed easy to handle by new users, but, at the same time, they are of great value because they are SAR raw data.

Level-1 are data intended for most users and they come into two different products:

- **Level-1 SLC (Single Look Complex):** focused SAR data geo-referenced using orbit and altitude data from the satellite. This product includes a single look in each dimension (range and azimuth) in complex format.
- **Level-1 GRD (Ground Range Detected):** focused SAR data that has been detected, multi-looked, and projected to ground range. Its multi-look nature reduces strongly the speckle content (but it reduces the spatial resolution). For that reason, three resolutions are available (depending on the amount of multi-looking performed), Full, High, and Medium.

Level-2: it has several special products for ocean observation:

- **OSW (Ocean Swell) spectra:** a two-dimensional ocean surface swell spectrum that includes an estimate of the wind speed and direction per swell spectrum. OSW operates in Stripmap and Wave modes.
- **OWI (Ocean Wind Fields):** intended for monitoring surface wind speed and direction at 10 m above the surface (it operates in Stripmap, Interferometric Wide Swath and, Extra Wide Swath mode modes).
- **RVL (Surface Radial Velocities):** a ground range gridded difference between the measured Level-2 Doppler grid and the Level-1 calculated geometrical Doppler. RVL provides an estimate of the width of the ocean Doppler spectra.

ESA has developed several software tools (toolboxes), freely available, for handling Sentinel data; SNAP (Sentinel Application Platform) is one of them. SNAP can be downloaded from the website https://step.esa.int/main/download/snap-download/, and a full description is available at https://step.esa.int/main/toolboxes/snap/. It is strongly recommended to use SNAP to process downloaded data from Copernicus Open Access Hub. SNAP's interface contains a link to many tutorials that are recommended to visit.

Figure 1.36 shows the graphical interface of the SNAP tool and a SAR image to be processed with it. We present examples of processing actual SAR data with SNAP in Section 1.9.

In the website https://asf.alaska.edu/how-to/data-recipes/how-to-create-a-subset-of-a-sentinel-1-product/, an actual Sentinel 1 scene is fully processed to get the final VV and VH polarized images.

1.8 NASA Earth Data Open Data

NASA's airborne and satellite systems provide also highly-valuable SAR data for users (many of them freely), although, sometimes it can be challenging to use. For that reason, NASA's Earth Science Data Systems (ESDS) program and Earth Observing System Data and Information System (EOSDIS) Distributed Active Archive Centers (DAACs) make available to users

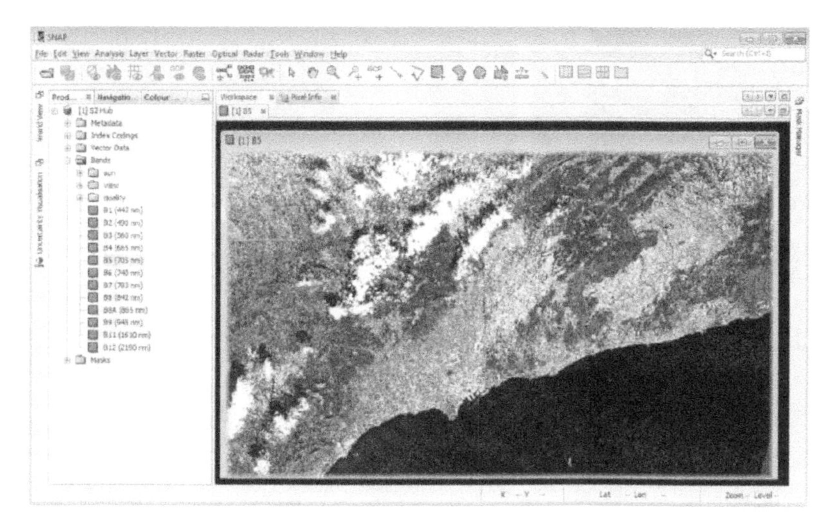

Figure 1.36 An example of processing SAR data with the SNAP tool.

a set of tools to overcome those difficulties. Users interested in accessing NASA's SAR data, are encouraged to visit the websites:

- https://earthdata.nasa.gov/learn/articles/getting-started-with-sar: provides an introduction to SAR and it details how to access to SAR data.
- https://worldview.earthdata.nasa.gov is a tool from NASA's Earth Observing System Data and Information System (EOSDIS) that provides the capability to interactively browse over 900 global, full-resolution satellite imagery layers and then download the underlying data.

1.9 Actual SAR Data Examples

This section is devoted to showing two actual SAR data sets (Sentinel-1) freely available. Data is downloaded from

- Hawaii's Big Island (4 equivalent looks): from NASA Earth Data Open Data (https://asf.alaska.edu/how-to/data-recipes/how-to-create-a-subset-of-a-sentinel-1-product/),
- Other examples: some Sentinel-1 actual SAR data.

The purpose is not to give a SNAP tutorial but to show some results of interest.

1.9.1 Hawaii's Big Island

Figure 1.37 shows the Intensity VH band over Hawaii's Big Island as it is available. The optical image from Google Earth is also shown just to point out a typical issue: the SAR geo-spatial image orientation does not correspond with the optical image (it is also mirrored, that is, the left part should be the right part and the opposite). The optical image has been rotated to note more easily the mirroring effect. To correct this orientation flaw is an trivial task (a simple image co-registration operation would solve it).

A subset of the original image is taken (in SNAP by using the utility "Spatial Subset from View..."), and the images for the Intensity and Amplitude bands in the polarization VH can

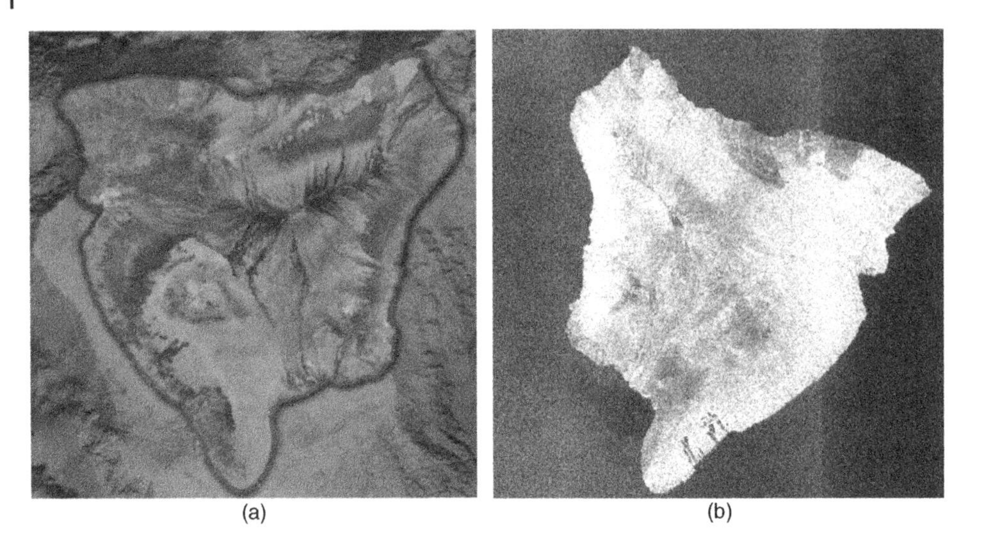

Figure 1.37 (a) Optical image from Google Earth of Hawaii's Big Island and the Sentinel-1 image (b).

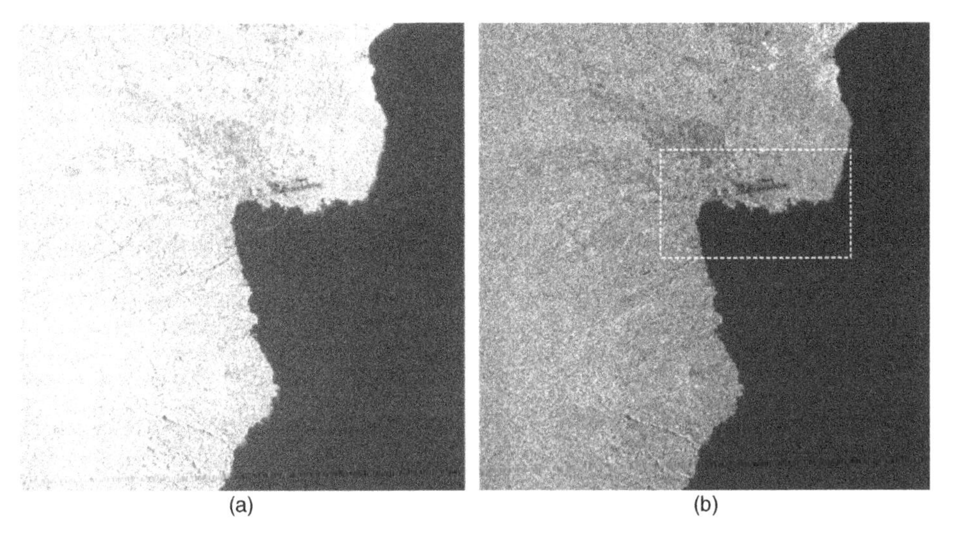

Figure 1.38 (a) Amplitude VH image and intensity VH image (b) for a subset from the original Sentinel-1 SAR data. See the next figure for a zoom of the dashed rectangle.

be seen in Figure 1.38. The differences between those data modalities, intensity and amplitude are addressed in Chapter 3, although, some clear differences are visible from the images.

From the many possibilities of the SNAP to work with satellite data, just some processing is done for the intensity SAR data. The common methodology includes:

- reading the data (already done). It is recommended not to uncompress the original downloaded data to save hard disk space. SNAP will handle the compressed data.
- the removal of the thermal noise from electrical fluctuations from the random thermal motion of electrons,
- steps for acquiring the satellite orbit file (to improve geo-coding and other capabilities),
- calibration to make that image pixels represent true radar backscatter values from the reflecting surface,

- terrain-flattenig to *flatten* the image, that is, to reduce the terrain-induced radiometric variations,
- terrain-correction step to apply a DEM (digital elevation model) to produce a map projected product,
- despeckling to reduce speckle content. SNAP includes the well-known Enhanced Lee filter mentioned above (Lee et al., 2009),
- write the transformed data to files.

Some of those steps are not required and sometimes other are depending of the nature of data (raw data, single-look-data, ground data, etc.) This done for the selected area shown in Figure 1.38 (the rectangle in dashed white lines). Figure 1.39 shows the processed image and its filtered version. As it can be seen, details are preserved, specially the bright scatters, while speckle has been reduced.

At last, SNAP can export the original SAR data bands (the data) and the processed bands in many formats. Among them, the one used in this book is the ENVI format (a file or files of binary data and a header file that collects all relevant information). This task can be done in SNAP as explained in Figure 1.40. The submenu "Bands extractor" opens from (Toolbar) Raster → Bands extractor.

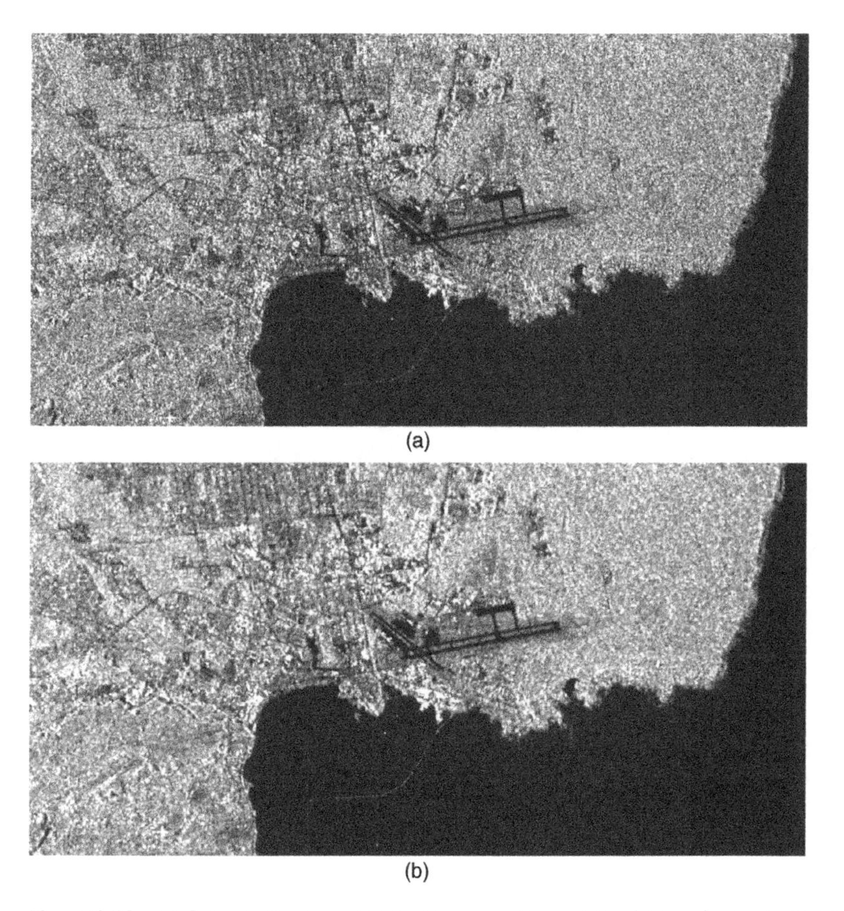

Figure 1.39 (a) Subset after applying some processing and the filtered image by the Lee filter(b).

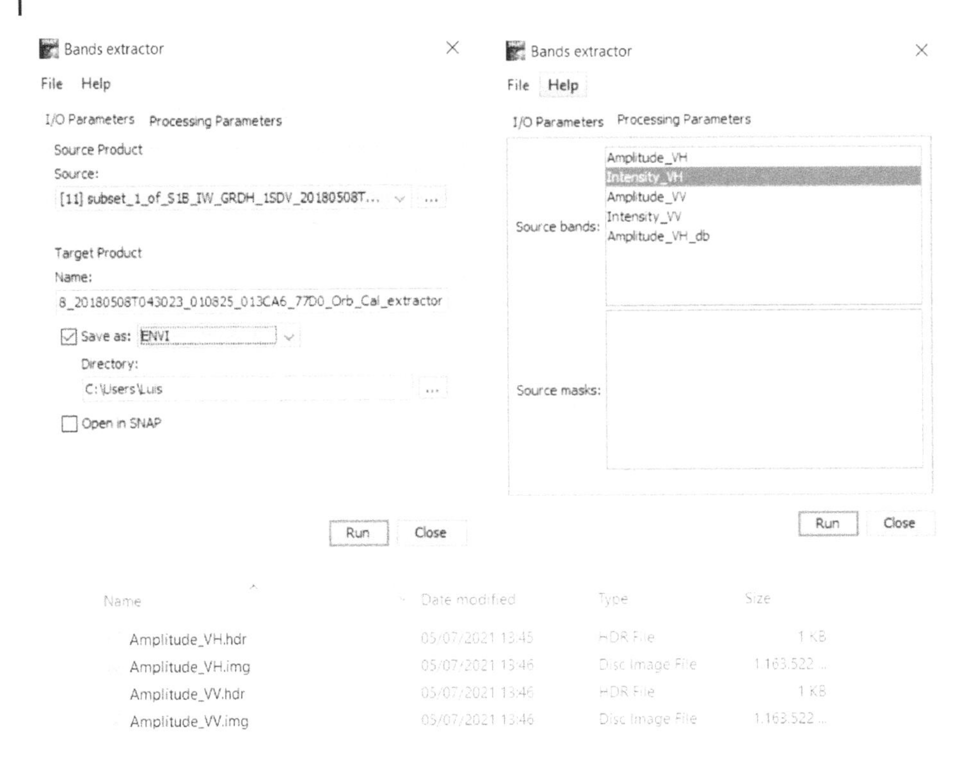

Figure 1.40 Exporting the bands from SNAP to ENVI format.

1.9.2 Other Examples

From the Copernicus Open Access HUB, some examples over Europe are selected and shown in Figure 1.41 and Figure 1.42.

Figure 1.41 Low-resolution Sentinel-1 intensity-HH image over Berlin and surroundings (image has been processed with SNAP).

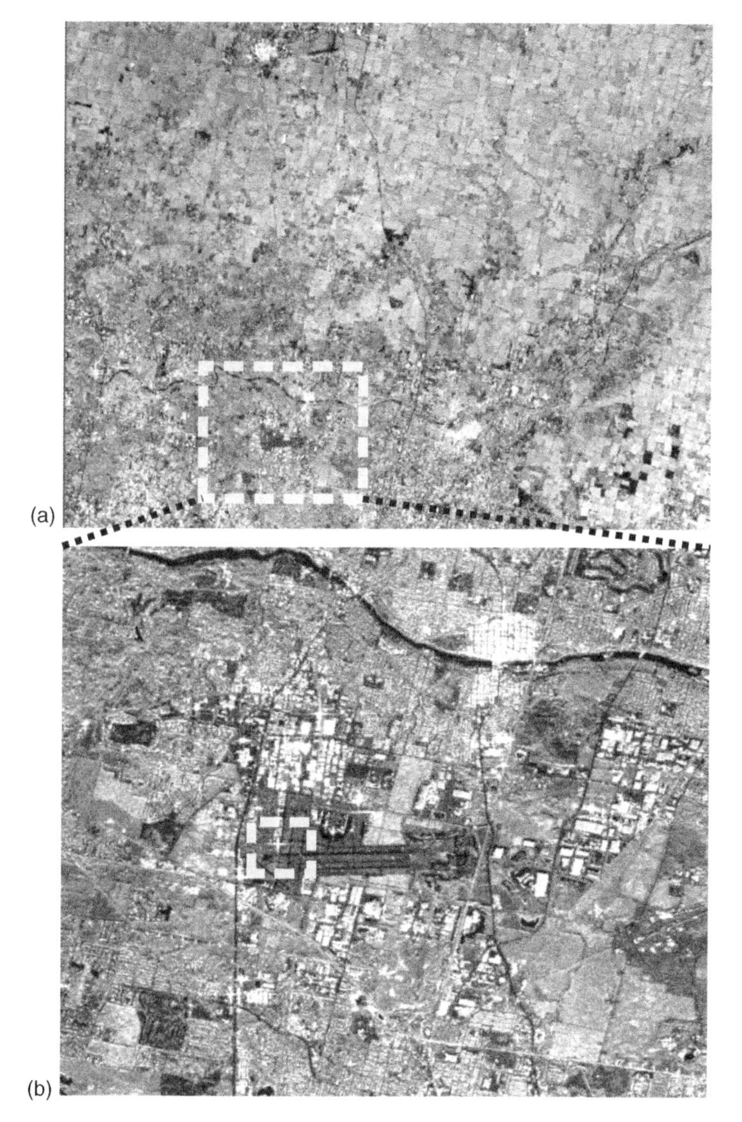

Figure 1.42 (a) High-resolution Sentinel-1 amplitude-HH image over Central Europe (image has been processed with SNAP). (b) The bottom image is a zoom of the indicated area, where a strong bright scatterer can be seen (within the dashed white rectangle).

Exercises

1 For a set of targets with backscatter coefficients, $\sigma_1 > \sigma_2 > \sigma_3$, and a given chirp-up radar signal, represent, in a one-dimensional plot, the received echoes.

2 In the assumption that the representation for Figure 1.33 is for an ascending-orbit satellite, represent the same for a descending-orbit satellite.

3 By using the Copernicus Open Access HUB, locate Sentinel-1 data and process them with the SNAP program.

2

Elements of Data Analysis and Image Processing with R

In this chapter, some fundamental background and tools about the data analysis will be given in detail, as the statistical properties of Synthetic Aperture Radar (SAR) data are extremely important for SAR image processing. In addition, due to the spatial property of SAR images, some operations about image processing are also introduced. To enable a greater understanding of the contents of this chapter, corresponding codes based on R are given.

2.1 Useful R Packages

It is well known that many useful R functions come in packages, free libraries of code written by R's active user community. To install an R package, open an R session and type at the command line:

```
install.packages("<the package's name>")
```

Note that, all packages of R are coming from CRAN (https://cran.r-project.org/). Thus, a connection to the internet is necessary for installing the packages. Once you have a package installed, you can make its contents available to use in your current R session by running:

```
library("<the package's name>")
```

Supported by R's community, there are thousands of helpful R packages for you to use. Since navigating them all is a challenge, some useful R packages are given in this section. These packages are also used in the book. In Wickham (2019), you can read about the entire package development process online. Also, using net(https://www.rdocumentation.org), you can do searching on the R packages. Note that, if the R language is installed in your computer, some R packages are also installed, such as `base`, `datasets`, `utils`, `grDevices`, `graphics`, `stats`, and `methods`. It means that you can use these packages without installation.

In Sections 2.1.1 and 2.1.2, some R packages used for data loading and manipulation are introduced. Note that, as an image can be considered as data of no less than two dimensions, the packages used for image loading and manipulation are also included.

2.1.1 Data Loading

For data analysis, a smooth loading of the data to be analyzed is the first step. Given the different formats or storage of data, there are some different packages for data loading. They are

SAR Image Analysis — A Computational Statistics Approach: With R Code, Data, and Applications, First Edition.
Alejandro C. Frery, Jie Wu, and Luis Gomez.

Companion website: www.wiley.com/go/frery/sarimageanalysis

- **xlsx** - These packages can help you read and write Microsoft Excel files. The corresponding R codes (available in the file `Code/R/Chapter2/211a.R`, see the contents of www.wiley.com/go/frery/sarimageanalysis) are:

```
library(xlsx)
rm(list = ls())
data <- c(1, 2, 3, 4, 5, 6, 7, 8)
write.xlsx(data, '../../../Data/CSV/test.xls', row.names = FALSE)
input <- read.xlsx('../../../Data/CSV/test.xls', sheetIndex = 1)
print(input)
```

Note that, R can handle plain text files by default. This means that using the functions, e.g. `read.csv`, `read.table`, `write.csv`, and `write.table`, you can import and export data from plain text files.

- **readbitmap** - This package can help you read image files of several formats, namely, `jpeg`, `png`, `tiff`. And, the input image is stored as an array. So, you can do some computation defined by yourself with the help of this package. The corresponding R code (available in the file `Code/R/Chapter2/211b.R` from www.wiley.com/go/frery/sarimageanalysis) is:

```
library(readbitmap)
rm(list = ls())
img <- readbitmap::read.bitmap('../../../Data/IMG/PNG/lena.png')
R <- nrow(img) / ncol(img)
plot(
        c(0, 1),
        c(0, r),
        type = "n",
        xlab = "",
        ylab = "",
        asp = 1)
rasterImage(img, 0, 0, 1, r)
```

- **imager** - This package can also help you read image files of several formats like that of the package `readbitmap`. However, the input image is stored as an object of `Cimg`. The corresponding R code (available in the file `Code/R/Chapter2/211c.R` from www.wiley.com/go/frery/sarimageanalysis) is:

```
library(imager)
rm(list = ls())
img <- load.image('../../../Data/IMG/PNG/lena.png')
plot(img,
 depth(img),
 xlim = c(1, width(img)),
 ylim = c(height(img), 1))
```

2.1.2 Data Manipulation

To facilitate a better data analysis, some data operations are necessary, including rearranging, reshaping, or joining together data sets. The corresponding R packages are

- **plyr** - This package can help you to do "groupwise" operations with your data, such as subsetting, summarizing, rearranging, and joining together data sets. One of the sample R codes (available in the file `Code/R/Chapter2/212a.R` from www.wiley.com/go/frery/sarimageanalysis) for part extraction is:

```
library(plyr)
rm(list=ls())
x <- array(seq_len(3 * 4 * 5), c(3, 4, 5))
```

```
print("**********************")
output <- take(x, 3, 1)
print(output)
print("**********************")
output <- take(x, 2, 1)
print(output)
print("**********************")
output <- take(x, 1, 1)
print(output)
print("**********************")
output <- take(x, 3, 1, drop = TRUE)
print(output)
print("**********************")
output <- take(x, 2, 1, drop = TRUE)
print(output)
print("**********************")
output <- take(x, 1, 1, drop = TRUE)
print(output)
print("**********************")
```

- **reshape2** - This package contains tools used for changing the layout of your data sets as the layout `R` likes best. One of the sample `R` codes (available in the file `Code/R/Chapter2/212b.R` from www.wiley.com/go/frery/sarimageanalysis) for using `melt()` is:

```
library(reshape2)
rm(list = ls())
print("**********************")
a <- array(c(1:23, NA), c(2, 3, 4))
b <- melt(a)
print(b)
print("**********************")
b <- melt(a, na.rm = TRUE)
print(b)
print("**********************")
b <- melt(a, varnames = c("X", "Y", "Z"))
print(b)
print("**********************")
```

- **imager** - This package contains some useful functions for image processing, such as rotation, blur, resize, and color space transform. One of the sample `R` codes (available in file `Code/R/Chapter2/212c.R` from www.wiley.com/go/frery/sarimageanalysis) for image resize is:

```
library(imager)
library(tiff)
rm(list = ls())
img <- readTIFF("../../../Data/IMG/TIFF/demoSAR.tiff")
resizeImg <- imresize(as.cimg(img), 1 / 4)
writeTIFF(as.matrix(resizeImg),
 "../../../Data/IMG/TIFF/resizeImg.tiff")
```

2.2 Descriptive Statistics

If no knowledge about the data is given, we need some tools to make exploration about the properties of data. This is very helpful for selecting a suitable model or method for data analysis. Since SAR data are numerical, some simple and useful descriptive statistical quantities are given in this section, which are widely used for SAR data analysis. To facilitate a good understanding of these quantities, they are introduced according to their different functions in Sections 2.2.1, 2.2.2, and 2.2.3. Note that, for a better explanation, an actual SAR image (shown in Figure 2.1) is adopted in the section.

Figure 2.1 An example of SAR imagery.

2.2.1 Center Tendency of Data

When comparing two datasets, the central location of the two data sets is very important. In statistics, there are several quantities used to describe the center tendency of the dataset: the mean, the median, and the mode.

Mean: It is a commonly used quantity to describe the average level of a data set. Assuming the observed data are $x_1, x_2, x_3, \dots, x_n$, it is defined as

$$\bar{x} = \frac{1}{n}\sum_{i=1}^{n} x_i = \frac{x_1 + x_2 + x_3 + \cdots + x_n}{n}, \tag{2.1}$$

where n is the number of samples contained in the dataset, and x_i is a sample. Obviously, an averaging operation is used to compute the **Mean**. If taking into account the frequency of occurrence of different values, a weighted averaging method is obtained to compute the **Mean**. The definition is

$$\bar{x} = \frac{w_1 x_1 + w_2 x_2 + w_3 x_3 + \cdots + w_n x_n}{w_1 + w_2 + w_3 + \cdots + w_n}, \tag{2.2}$$

where n is the number of samples of different values contained in the dataset, w_i is the frequency of occurrence corresponding to x_i.

Median: If there are some abnormal samples in the dataset, the **Mean** will not be suitable to be used as the central location. Under this condition, the **Median** is a good replacement for the description of central tendency. Supposing all the samples in the dataset are sorted in ascending or descending order as $x_{(1)}, x_{(2)}, x_{(3)}, \dots, x_{(n)}$, the **Median** is defined as:

$$M_e = \begin{cases} x_{\lfloor \frac{n+1}{2} \rfloor}, & \text{if } n \text{ is odd} \\ \frac{x_{\lfloor \frac{n}{2} \rfloor} + x_{\lfloor \frac{n}{2} \rfloor + 1}}{2}, & \text{if } n \text{ is even,} \end{cases} \tag{2.3}$$

where $\lfloor \cdot \rfloor$ is the function *floor*$(\cdot)$. From the definition, it will be found that only the order information is used to compute the **Median**. Thus, the **Median** is a robust estimator (Maronna et al., 2006).

Mode: Another quantity for describing the center tendency, the **Mode** is computed as the value that is of the highest frequency. This quantity may not be defined (as in the case of a perfectly uniform sample), or a sample may have multiple modes.

α-quantile: As a quantity used to describe the relative position in a dataset, the **α -quantile** (e.g. q_α) is defined as

$$q_\alpha = \inf \{t : F_x(t) \geq \alpha\}, \tag{2.4}$$

where $F_x(\cdot)$ is the dataset empirical cumulative distribution function. So, the **Median** is $q_{1/2}$.

Using actual SAR imagery, as the one shown in Figure 2.1, the aforementioned quantities can be computed via the following R code (available in the file Code/R/Chapter2/221.R from www.wiley.com/go/frery/sarimageanalysis):

```
library(tiff)
rm(list = ls())
img <- readTIFF("../../../Data/IMG/TIFF/demoSAR.tiff")
# computing mean value
meVal <- mean(img)
print(meVal)
# computing mode value
imghist <- hist(img)
maxval <- max(imghist$counts)
maxind <- which(imghist$counts == maxval)
mode <- imghist$mids[maxind]
print(mode)
# computing median value
medval <- median(img)
print(medval)
# computing alpha value
alpha = 0.5
cumfreq <- rep(1 / length(img), length(img))
cumfreq <- cumsum(cumfreq)
alphaInd <- min(which(cumfreq >= alpha))
sortData <- sort(img)
print(sortData[alphaInd])
```

More generally, and regardless of the sample size, the sample quantile of order $\alpha \in (1/n, 1 - 1/n)$ is a weighted average of consecutive order statistics. Hyndman and Fan (1996) provide a comprehensive account of how several computational platforms implement sample quantiles.

2.2.2 Dispersion of Data

With the help of quantities defined in Section 2.2.1, only the central location can be calculated. For the spread of the sample around the center location, some other quantities are needed, such as the **Range**, the **Variance**, and the **Coefficient of Variation**. By using these kinds of quantities, we can get more information about the shape of a dataset.

Range: The definition of **Range** is the absolute difference between the minimum and maximum values in a dataset. Usually, a larger value of **Range** implies a higher dispersion of the dataset. Similarly, we can use the absolute difference between $q_{1/25}$ and $q_{1/75}$ to measure the spread of a dataset, which is defined as **Interquantile Range**.

Variance: Since only two samples' values are used to compute the **Range**, the information it provides is very limited. So, it is more popular to use the **Variance** as a measure of dispersion. Assuming the observed samples are $x_1, x_2, x_3, \dots, x_n$, the sample's variance is defined as

$$\begin{aligned} s^2 &= \frac{1}{n-1}\sum_{i=1}^{n}(x_i - \overline{x})^2 \\ &= \frac{(x_1-\overline{x})^2 + (x_2-\overline{x})^2 + (x_3-\overline{x})^2 + \cdots + (x_n-\overline{x})^2}{n-1}, \end{aligned} \tag{2.5}$$

where n is the number of samples and $\overline{x}$ is computed by Eq. (2.1). Intuitively, Eq. (2.5) is also an averaging operation. Thus, according to Eq. (2.2) and using w_i as the frequency corresponding to x_i, the weighted variance is computed as

$$\begin{aligned} s^2 &= \frac{1}{n-1}\sum_{i=1}^{n}w_i(x_i - \overline{x})^2 \\ &= \frac{w_1(x_1-\overline{x})^2 + w_2(x_2-\overline{x})^2 + \cdots + w_n(x_n-\overline{x})^2}{\sum_{i=1}^{n} w_i - 1}. \end{aligned} \tag{2.6}$$

Usually, a larger **Variance** means a greater spread of the data. However, due to the usage of a square operation, the unit of **Variance** is not same as that of the original data. Thus, the sample standard deviation, that is in the same units as the data, is always used to measure the spreading of data. The definition of the sample standard deviation is

$$s = \sqrt{s^2} \tag{2.7}$$

Coefficient of Variance (CoV): Also denoted CV, is the ratio of the sample standard deviation to the sample mean:

$$\text{CoV} = \frac{s}{\overline{x}}, \tag{2.8}$$

where $\overline{x}$ is computed by Eq. (2.1). Due to the use of a ratio operation, CoVs of different datasets can be compared directly. Note that, the condition for the use of CoV is that the sample's value of the dataset should be positive to avoid division by zero.

Now we compute the aforementioned values using an actual SAR image via the following R code (available in the file `Code/R/Chapter2/222.R` from www.wiley.com/go/frery/sarimageanalysis):

```
rm(list = ls())
library(tiff)
img <- readTIFF("../../../Data/IMG/TIFF/demoSAR.tiff")
# computing range
source("rangeVal.R")
rang <- rangeVal(img, 0.25, 0.75)
print(rang)
# computing sample's variance
samvar <- sum((img - mean(img)) ^ 2) / (length(img) - 1)
```

```
print(samvar)
# computing cov
covval <- sqrt(samvar) / mean(img)
print(covval)
```

The function of rangeVal is defined as:

```
rangeVal <- function(data, alpha, beta) {
        cumfreq <- rep(1 / length(data), length(data))
        cumfreq <- cumsum(cumfreq)
        alphaInd <- min(which(cumfreq >= alpha))
        betaInd <- min(which(cumfreq >= beta))
        sortData <- sort(data)
        val <- abs(sortData[alphaInd] - sortData[betaInd])
        return(val)
}
```

Note that, in the `R` language, using functions `range()`, `sd()`, and `quantile()`, the range, standard deviation, and quantile values of the data can also be calculated. The interquartile range is the difference $\text{IQR}(\boldsymbol{x}) = q_{3/4}(\boldsymbol{x}) - q_{1/4}(\boldsymbol{x})$.

2.2.3 Shape of Data

To make possible a further description of the statistics of the data, some specialized quantities are designed to reflect the shape of the data's distribution. They are skewness and kurtosis. Essentially, they are defined with the help of high-order statistics.

Skewness: This quantity was proposed by Pearson (1895) to measure the symmetry (or asymmetry thereof) of the data distribution. To compute **Skewness** (S_k), it is defined as

$$S_k = \frac{n\sum_{i=1}^{n}(x_i - \overline{x})^3}{(n-1)(n-2)s^3}. \tag{2.9}$$

From this definition, it can be seen that S_k is computed via third-order statistics and is a signed quantity. The sample distribution is symmetric when $S_k = 0$. When $S_k < 0$, the distribution is left-skewed, while it is right-skewed if $S_k > 0$.

Kurtosis: This quantity was first proposed by Pearson (1905) and used to measure the steepness of a distribution. It is computed as

$$K_u = \frac{n(n+1)\sum_{i=1}^{n}(x_i - \overline{x})^4 - 3(n-1)^3 s^4}{(n-1)(n-2)(n-3)s^4}. \tag{2.10}$$

From this definition, we see that K_u is computed using fourth-order statistics. Usually, when the distribution of data approximates normal, the value of K_u will approximate 0. When $K_u < 0$, it means the shape of the distribution is flatter than that of a Normal distribution, while it is steeper than that of Normal distribution when $K_u > 0$.

Thus, using SAR data (shown in Figure 2.1), these two shape quantities can be computed via the following `R` code (available in file `Code/R/Chapter2/223.R` from www.wiley.com/go/frery/sarimageanalysis).

```
rm(list = ls())
library(tiff)
img <- readTIFF("../../../Data/IMG/TIFF/demoSAR.tiff")
# computing skewness
n <- length(img)
tval <- sum((img - mean(img)) ^ 3)
skval <- (n * tval) / ((n - 1) * (n - 2) * sd(img) ^ 3)
print(skval)
# computing Kurtosis
```

```
n <- length(img)
tval <- sum((img - mean(img)) ^ 4)
val0 <- n * (n + 1) * tval - 3 * (n - 1) ^ 3 * sd(img) ^ 4
val1 <- (n - 1) * (n - 2) * (n - 3) * sd(img) ^ 4
kurval <- val0 / val1
print(kurval)
```

Note that, in the R language, using functions `skewness()` and `kurtosis()`, the values corresponding to skewness and kurtosis of the data can also be calculated. It is always advisable to use the native R functions, rather than implementing them from scratch.

2.3 Visualization

In statistics, using a visualization of the data, we can obtain a preliminary perception of the properties about the distribution that produced the measurements. Compared with the descriptive quantities shown in Section 2.2, data visualization is very important at showing a full scene about the distribution of data whose dimension is no larger than three. With this preliminary perception, we can make good decisions about which statistical models to use to fit the observed samples. Note that, a good data model is key for a further deduction based on the observed samples.

According to the type of data (e.g. categorical or numerical) or the target of data analysis, there are many different visualization tools, such as histograms, radar charts, box plots, scatter diagrams, and stem-and-leaf displays.

Since our book concentrates on the data analysis of SAR imagery, histograms (for one dimension data) and scatter diagrams (for two-dimensional data) are introduced after the introduction of rug and box plots in this section.

2.3.1 Rug and Box Plots

As the simplest graphical display of a univariate sample, "rug plot" consists of marks corresponding to each observation along an axis. Note that, rug plots are seldom useful alone; they complement the information provided by the two most commonly seen displays of univariate samples: the boxplot (and variations), and the histogram (and its smoothed versions).

The classical "in the style of Tukey" boxplot shows five summary statistics (the median, two hinges and two whiskers), and all "outlying" points individually. The hinges are the first and third quartile. The right whisker is set as the smallest value that is no less than $q_{3/4}(\boldsymbol{x}) + \frac{1}{2}\mathrm{IQR}(\boldsymbol{x})$. The left whisker is set as the largest value that is no larger than $q_{1/4}(\boldsymbol{x}) - \frac{1}{2}\mathrm{IQR}(\boldsymbol{x})$. Here, $q(\cdot)$ is the quartile function defined in Eq. 2.4. Besides, it is useful to plot a notch around the median, usually of size $1.58n^{-1/2}\mathrm{IQR}(\boldsymbol{x})$ that corresponds to approximately a 95 % symmetric confidence interval. Outlying points are above the right whisker and below the left whisker.

Figure 2.2 shows the elements of a boxplot. The observations are mapped onto the rug, in the abscissas axis. The only outlying observation is shown as an isolated dot. Notice that the horizontal grid has been omitted, as it carries no useful information in this plot and may be misleading.

Boxplots are particularly useful when comparing two or more samples. Figure 2.3 shows the boxplots of three samples. They have approximately the same sample mean values, but the spread of the data is noticeable different.

While boxplots are built with summaries based on the sorted sample, the histogram and its variations employ all the data. As such, the rug is a histogram. More often than not,

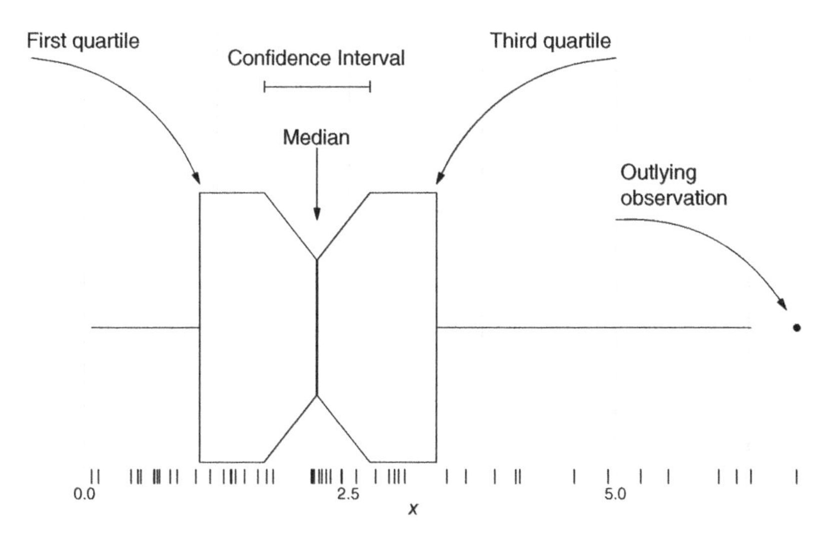

Figure 2.2 Elements of a boxplot with rug.

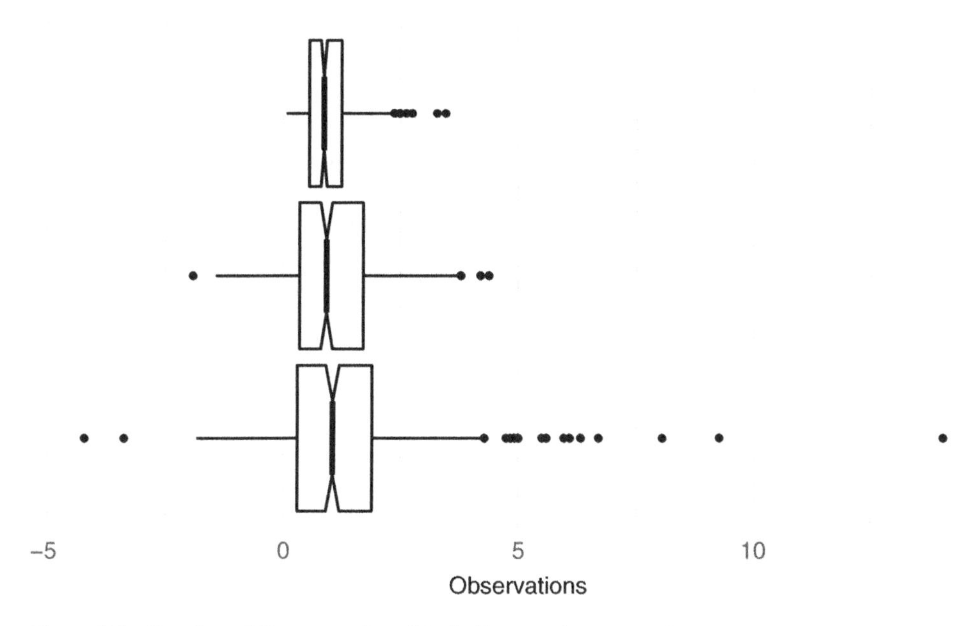

Figure 2.3 Boxplots of three samples with similar sample mean values.

histograms show the counts of observations within disjoint intervals ("bins"), instead of marking each observation as in a rug.

2.3.2 Histogram

Using the random sampling method, a dataset of 1,000 samples is constructed from the Normal distribution $\mathcal{N}(1,1)$ whose histogram is shown in Figure 2.4(a). Note that, the height of each bin is computed as:

$$\widehat{F}(t) = \frac{1}{n}\#\{j : |z_j - b_t| \leq b_{hw}\}, \tag{2.11}$$

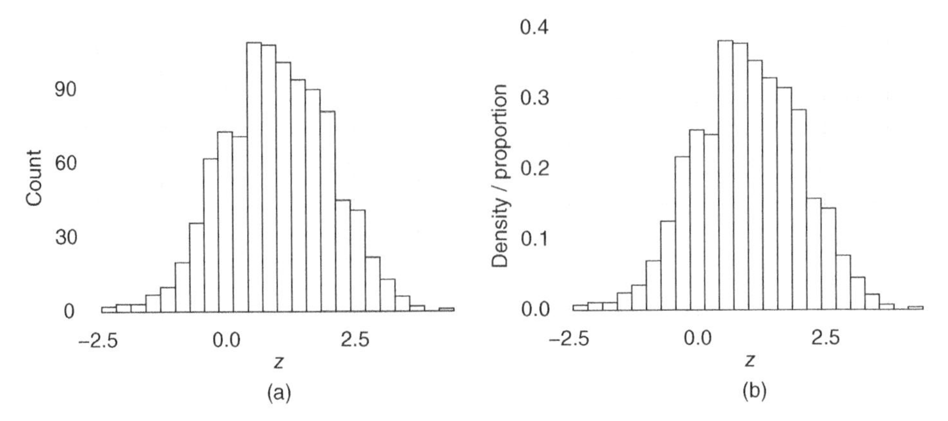

Figure 2.4 Histograms of data sampled from **N**(1,1). (a) Histogram of counts and (b) histogram of proportions.

where t is the index of the bin whose center value is b_t, and b_{hw} is the half-width of each bin in the histogram. The graphical display of (2.11) yields the histogram of counts shown in Figure 2.4a.

If we divide each count by the total number of observations:

$$\hat{p}(t) = \hat{F}/(2 \times b_{hw}), \tag{2.12}$$

we obtain the histogram of proportions; cf. Fig 2.4b. Eq (2.12) is an estimator of the underlying density which characterizes the distribution that produced the data.

Different choices of b_{hw} produce different histograms. Freedman and Diaconis (1981) discuss procedures whereby Eq. (2.12) converges to the true density. A simple rule-of-the-thumb, known as "Freedman-Diaconis rule" for such a choice is selecting b_{hw} such that the number of bins equals $2\text{IQR}/n^3$, where IQR denotes the inter-quartile range.

Moreover, given the definition of the cumulative distribution function (CDF), the empirical CDF is computed as:

$$\hat{F}(t) = \frac{1}{n}\#\{j : z_j \leq t\}. \tag{2.13}$$

The empirical CDF of the data, whose histogram of proportions is shown in Figure 2.4b, is shown in Figure 2.5a. The non-decreasing property of CDF is observed by the empirical CDF, i.e. the empirical cumulative distribution function is a cumulative distribution function.

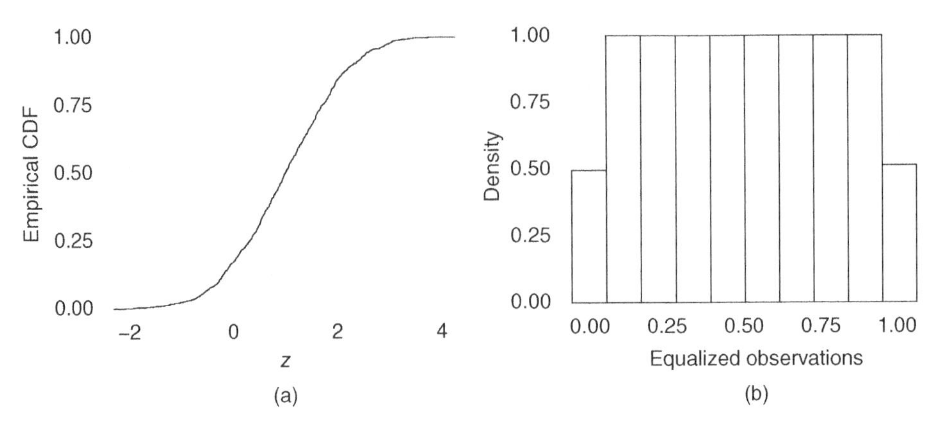

Figure 2.5 Empirical CDF and equalized histograms corresponding to the data shown in Figure 2.4. (a) Empirical cumulative distribution function and (b) histogram of the equalized image.

Theorem 2.1 Consider the continuous random variable $Z : \Omega \to \mathbb{R}$ with CDF F_Z. The random variable $W = F_Z(z)$ has uniform distribution.

Proof: The CDF of W is $F_W(w) = \Pr(W \leq w) = \Pr(F_Z(z) \leq w)$. Since Z is continuous, there exists the inverse of its CDF, F_Z^{-1}. We can then write $F_W(w) = \Pr(F_Z(z) \leq w) = \Pr(F_Z^{-1}(F_Z(z)) \leq F_Z^{-1}(w)) = \Pr(Z \leq F_Z^{-1}(w))$, which is exactly the CDF of Z at $F_Z^{-1}(w)$, so $F_W(w) = F_Z(F_Z^{-1}(w)) = w$ is a uniform random variable on (0,1).

From theorem 2.1, we can see that, using the CDF of a random variable, the random variable will be mapped into a random variable obeying the Uniform distribution. Thus, using the sample cumulative distribution function on the samples that produced it, we will obtain samples coming from a distribution that is approximately the Uniform distribution. This observation leads to the histogram equalization technique. Deviations from the theoretical distribution are due to the fact that we are working with a finite sample. In Figure 2.5(b), an equalized histogram corresponding to Figure 2.4(a) is given. The main `R` codes used to produce Figures 2.4 and 2.5 (available in file `Code/R/Chapter2/231.R` from www.wiley.com/go/frery/sarimageanalysis) are the following:

```
rm(list = ls())
library("ggplot2")
set.seed(14012021)
Z <- rnorm(1000, 1)
W <- ecdf(Z)(Z)
HistogramEqualization = data.frame(Z, W)

ggplot(data = HistogramEqualization, aes(x = Z)) +
 geom_histogram(aes(y =..count..), col = "black", fill = "white") +
 xlab("z") +
 ylab("Count") +
 theme(text = element_text(size = 20))
ggplot(data = HistogramEqualization, aes(x = Z)) +
 geom_histogram(aes(y =..density..), col = "black", fill = "white") +
 xlab("z") +
 ylab("Density") +
 theme(text = element_text(size = 20))
ggplot(data = HistogramEqualization, aes(x = Z)) +
 stat_ecdf(geom = "step", pad = FALSE) +
 xlab("z") +
 ylab("Empirical CDF") +
 theme(text = element_text(size = 20))
ggplot(data = HistogramEqualization, aes(x = W)) +
 geom_histogram(aes(y =..density..), col = "black", fill = "white") +
 xlab("Equalized Observations") +
 ylab("Density") +
 theme(text = element_text(size = 20))
```

2.3.3 Scattering Diagram

Obviously, using the histogram, the randomness of the one-dimensional data can be well visualized. However, for data of no less than two dimensions, the visualization becomes more complex. Here, using two-dimensional Gaussian data as an example, the scattering diagram is given in Figure 2.6. From Figure 2.6, the statistical distributions of two two-dimensional Gaussian datasets are displayed. Note that, the distributions of the two-dimension Gaussian data are:

$$Data_1 \sim \mathbf{N}\left(\begin{bmatrix}1\\1\end{bmatrix};\begin{bmatrix}0.5 & 0\\0 & 0.5\end{bmatrix}\right) \tag{2.14}$$

$$Data_2 \sim \mathbf{N}\left(\begin{bmatrix}3\\3\end{bmatrix};\begin{bmatrix}0.5 & 0\\0 & 0.5\end{bmatrix}\right). \tag{2.15}$$

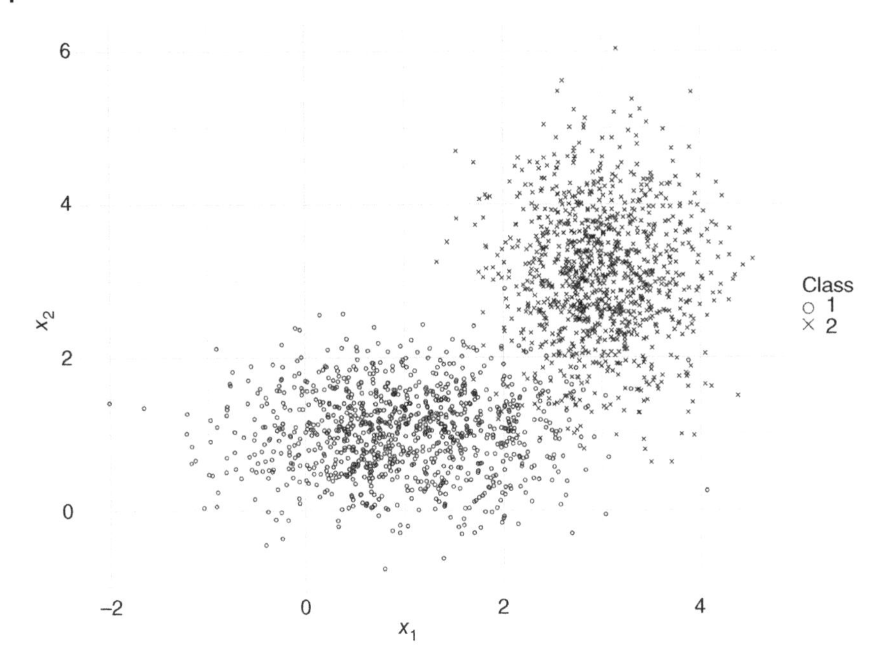

Figure 2.6 Scattering diagram of the samples produced by the distributions specified in Eqs. (2.14) and (2.15).

The corresponding R codes (available in file `Code/R/Chapter2/232.R` from www.wiley.com/go/frery/sarimageanalysis) are as following:

```
rm(list = ls())
library(mvtnorm)
library(latex2exp)
set.seed(123456789, kind="Mersenne-Twister")

sigma0 <- matrix(c(0.7, 0, 0, 0.3), nrow = 2)
data0 <- rmvnorm(1000, mean = c(1, 1), sigma0)
sigma1 <- matrix(c(0.3, 0, 0, 0.7), nrow = 2)
data1 <- rmvnorm(1000, mean = c(3, 3), sigma1)
df <- data.frame(rbind(data0, data1))
df$Class <- as.factor(rep(c(1, 2), c(1000, 1000)))

ggplot(df, aes(x = X1, y = X2, shape=Class)) +
        geom_point(alpha =.8) +
        scale_shape_manual(values = c(1, 4)) +
        xlab(expression(italic(x) [1])) +
        ylab(expression(italic(x) [2])) +
        guides(shape = guide_legend(override.aes = list(size = 5,
                alpha=1))) +
        theme_minimal() +
        theme(text=element_text(size=40, family="Times New Roman"))
```

Note that, the use of packages `mvtnrom` is for producing multidimensional normal variates.

2.4 Statistics and Image Processing

It is well known that, by ignoring the spatial and the band information contained in an image, an image can be considered as an array of values. Such values can, then, be mapped onto gray levels by a look-up table. So, the quantities (given in Section 2.2) and the visualization tools (described in Section 2.3) can be directly applied to explore the randomness of the image. In this section, using the aforementioned methods, some visual results are given in details.

2.4.1 Histogram-Based Image Transformation

In practice, to better visualize an image, some enhancing techniques are needed for image processing. As a simple method, histogram equalization is usually used for the image enhancement, in which the rate of different gray values is adopted for equalization (determining the equalized value).

However, in practice, one may find that equalized images are a bit unrealistic. There is another theorem that nicely complements Theorem 2.1 and yields any possible distribution from a uniform law. We will need the notion of *generalized inverse functions* to deal equally with continuous and discrete random variables.

Assume $g : \mathbb{R} \to \mathbb{R}$ is a nondecreasing function, e.g. a cumulative distribution function. If g is continuous and strictly increasing it has an inverse g^{-1}. Otherwise, we define its generalized inverse function as $g^{-}(y) = \inf \{x \in \mathbb{R} : g(x) \geq y\}$. If g is continuous and strictly increasing, g^{-1} and g^{-} coincide.

Theorem 2.2 ***(Inversion Theorem)*** Consider the random variable U uniformly distributed in (0,1), and the random variable V whose distribution is characterized by the cumulative distribution function F_V. The random variable defined by $Y = F_V^{-}(U)$ has the same distribution of V.

Proof: The distribution of Y is characterized by its cumulative distribution function $F_Y(t) = \Pr(Y \leq t)$. But $\Pr(Y \leq t) = \Pr(F_V^{-}(U) \leq t)$, and applying F_V to both sides of the inequality we have $\Pr\left(F_V(F_V^{-}(U)) \leq F_V(t)\right) = \Pr(U \leq F_V(t))$. Since U is uniformly distributed over the unit interval, $\Pr(U \leq F_V(t)) = F_V(t)$, so $F_Y(t) = F_V(t)$.

Theorems 2.1 and 2.2 provide powerful tools to stipulate (approximately) the marginal distribution of data. Schematically, if we have observations from Z and we want to transform them into observations of Y, the procedure consists of first estimating F_Z by $\widehat{F}_Z$, and applying it to the original data. These data will then be approximately uniform, then we apply F_Y^{-} to the data. Then, the new observations will have the desired histogram.

Notice that F_Y does not need to be specified theoretically. It can be also computed from data. With this, we can match the histogram of one image to that of another image; this operation is called "histogram matching," and it is useful for making two data sets visually comparable.

It is also important to notice that if F is a cumulative distribution function, then F^{-} is called a *quantile function*.

Here, we use the SAR image shown in Figure 2.1 as an example, which is an urban area. This means it has extreme variability. Figure 2.7a shows the histogram of the original data, while the stipulated histogram with Beta distribution ($\mathcal{B}(8,8)$) is given in Figure 2.7b. Note that the Beta distribution is a convenient choice since it has compact support and it is flexible enough to allow stipulating several shapes. Also notice that there are saturated values in the original image (the large rightmost bin) that are preserved in the transformation. In practice,

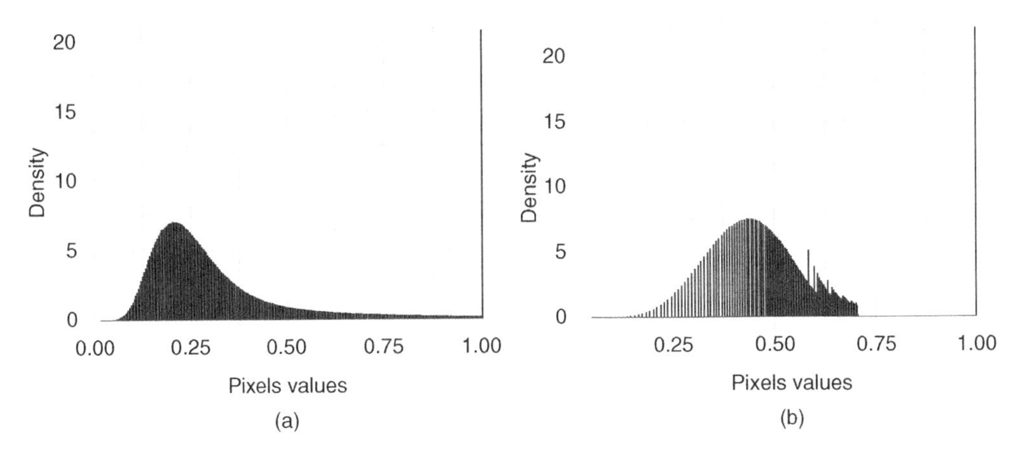

Figure 2.7 Histograms of the original and Beta based transformed data from an urban area. (a) Original image and (b) Image after histogram matching.

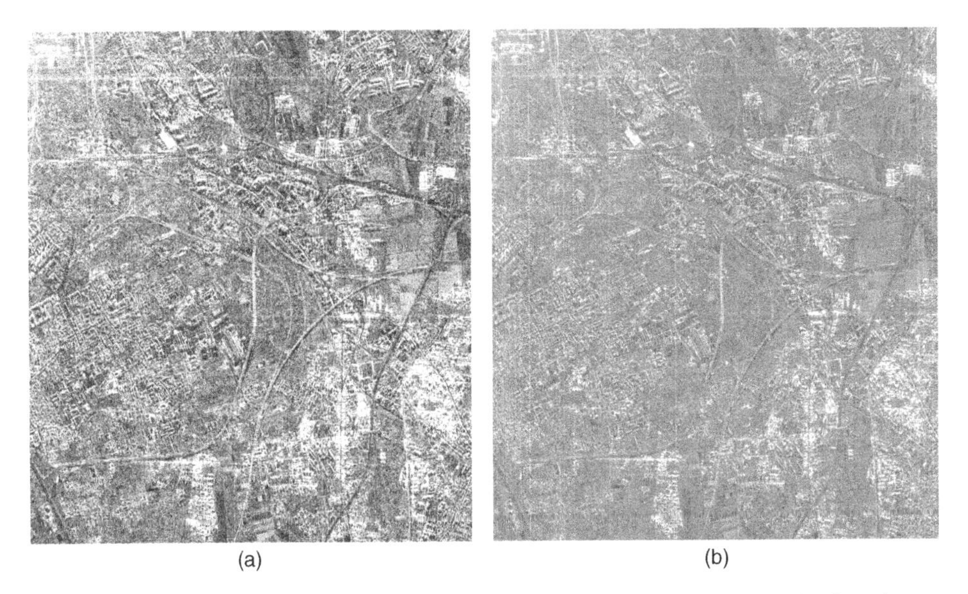

Figure 2.8 Transformed SAR images by histogram equalization and Beta distribution based modification. (a) Histogram-equalized image and (b) histogram-matched image.

this transformation does not require altering the data; it would suffice to build a new look-up table that maps values into gray levels.

Figure 2.8 shows the image after histogram equalization and after matching the histogram with a Beta distribution. The result obtained with Beta distribution-based transformation is less dramatic than the equalized version.

The corresponding `R` codes (available in file `Code/R/Chapter2/241.R` from www.wiley.com/go/frery/sarimageanalysis) used to obtain the results shown in Figures 2.7 and 2.8 are given as follows:

```
rm(list = ls())
library(tiff)
library(ggplot2)
img <- readTIFF("../../../Data/IMG/TIFF/demoSAR.tiff")
```

```
dimVals <- dim(img)
# plotting histogram of original data
imgData <- as.vector(img)
df <- data.frame(imgData)
ggplot(df, aes(x = imgData)) +
 geom_histogram(aes(y =..density..), col = "black", fill = "white") +
 xlab("pixel's values") +
 ylab("Density") +
 theme(text = element_text(size = 20))
# plotting histogram of transformed data with equalization
bright_imgData <- qunif(ecdf(imgData)(imgData))
df <- data.frame(bright_imgData)
ggplot(df, aes(x = bright_imgData)) +
 geom_histogram(aes(y =..density..), col = "black", fill = "white") +
 xlab("pixel's values") +
 ylab("Density") +
 theme(text = element_text(size = 20))
img_trans <- matrix(bright_imgData, nrow = dimVals[1])
# saving transformed data as image
writeTIFF(img_trans, "../../../Data/IMG/TIFF/DemoSARTransUnif.tiff")
# plotting histogram of transformed data with beta distribution
bright_imgData <- qbeta(ecdf(imgData)(imgData), shape1 = 8, shape2 = 8)
df <- data.frame(bright_imgData)
ggplot(df, aes(x = bright_imgData)) +
 geom_histogram(aes(y =..density..), col = "black", fill = "white") +
 xlab("pixel's values") +
 ylab("Density") +
 theme(text = element_text(size = 20))
# saving transformed data as image
img_trans <- matrix(bright_imgData, nrow = dimVals[1])
writeTIFF(img_trans, "../../../Data/IMG/TIFF/DemoSARTransBeta.tiff")
```

Note that, the usage of package `tiff` is for reading and writing images.

2.4.2 Scattering based Analysis

In SAR image analysis, due to the differences between different scenes, the local variance is very important for fine analysis of scene details, especially for the estimation of the equivalent number of looks. Moreover, given the coherent imaging method used in a SAR system, the speckle model is very important for the further analysis of the SAR images. Assuming that the data are locally stationary, the relationship between the local mean and local standard deviation can be used for speckle model analysis. Here, using the SAR image shown in Figure 2.1, we use 4×4 blocks to compute the local mean and local standard deviation. Figure 2.9 shows the scatter diagram.

Visually, a parabolic curve is obtained in Figure 2.9. Here, we use the linear regression method to determine a straight line. With a careful analysis, most points locate close to the straight line. This means the relationship between the local mean and the local standard deviation can be approximately represented as a straight line. Thus, under the assumption of local stationarity, the multiplicative speckle model is valid in this SAR image. For the points lying to the right of the line in the figure, they come from the local region of higher backscatter values and lower variance (such as the buildings). The slope and the intercept of the straight line are, respectively, 0.33976 and 0.000255. It is clear that the intercept of the straight line is very small. Thus, the usage of the multiplicative speckle model is reasonable for this image. To improve the visualization, using the slope of the obtained straight line as the base, two straight lines obtained with the same intercept and the slope of 15-degree difference are also given and shown as a dotted-line. Obviously, most points lie in the range covered by these two dotted-lines. The corresponding `R` codes (available in file `Code/R/Chapter2/242.R` from www.wiley.com/go/frery/sarimageanalysis) are given as follows.

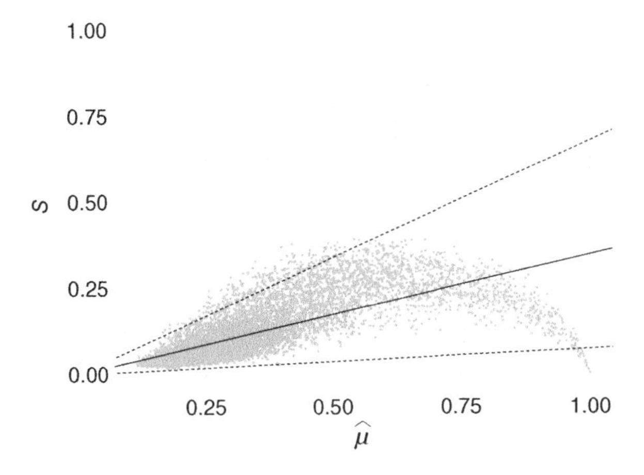

Figure 2.9 Scattering diagram of real SAR image between local mean and local standard variance.

```
rm(list = ls())
library(tiff)
library(ggplot2)
img <- readTIFF("../../../Data/IMG/TIFF/demoSAR.tiff")
dimVals <- dim(img)
step <- 4
numI <- dimVals[1] %/% step
numJ <- dimVals[2] %/% step
ni <- as.vector(seq.int(from = 1, to = dimVals[1], by = step))
nj <- as.vector(seq.int(from = 1, to = dimVals[2], by = step))
ni <- ni[1:numI]
nj <- nj[1:numJ]
ls <- 0:3
imgP <- array()
for (i in ls) {
 for (j in ls) {
  imgP <- rbind(imgP, as.vector(img[ni + i, nj + j]))
 }
}
imgP <- imgP[2:nrow(imgP),]
meVals <- apply(imgP, 2, mean)
sdVals <- apply(imgP, 2, sd)
index <- sample(1:length(meVals), 10000)
meVals <- meVals[index]
sdVals <- sdVals[index]
relation <- lm(sdVals ~ meVals)
# computing range of slp
slp <- relation$coefficients[2]
interp <- relation$coefficients[1]
deg <- atan(slp) * 180 / pi
deg_res <- 15
deg_rang <- deg + c(deg_res, -deg_res)
deg_rang <- deg_rang * pi / 180
slp_rang <- tan(deg_rang)
# plotting
df <- data.frame(meVals, sdVals)
ggplot(df, mapping = aes(x = meVals, y = sdVals)) +
 geom_point(color = "grey") +
 xlab("mu") +
 ylab("sigma") +
 ylim(c(0, 1)) +
 geom_abline(slope = slp,
  intercept = interp,
  linetype = 1) +
 geom_abline(slope = slp_rang[1],
  intercept = interp,
  linetype = 2) +
```

```
  geom_abline(slope = slp_rang[2],
   intercept = interp,
   linetype = 2) +
  theme() +
theme_bw()
```

We do not provide the estimate of the number of looks because it would be biased by the leftmost points. A deeper analysis of this estimation is discussed later in this book.

2.5 The imagematrix Package

There are a few packages for the R language that ease the visualization of images. Among them, imagematrix stands out for its simplicity. Although it has been deprecated and is no longer supported, we provide it with a few additional features. The source code of imagematrix is appended in Code/R/imagematrix/imagematrix.R. Note that, for the usage of imagematrix, the packages png and plot3D should be installed in advance.

The very basic rules for using imagematrix are:

- image data should be in the form of a matrix of m rows and n columns, with either 1 or 3 slices; the former will be rendered as a gray-levels image, the latter as a color image;
- the data must be in the [0,1] range, otherwise, it will be clipped.

Let us see an example (R codes are available in file Code/R/Chapter2/251.R from www.wiley.com/go/frery/sarimageanalysis).

```
rm(list = ls())
source("../imagematrix/imagematrix.R")
library(png)
require(plot3D)
x <- 1:500
y <- 1:100
xy <- mesh(x, y)
ramp <- (xy$x + xy$y) / 600
plot(imagematrix(ramp))
imagematrixPNG(imagematrix(ramp), "../../../Figures/PNG/chapter2/ramp.png")
```

The first line loads your local version of imagematrix, while the second makes the plot3D package available. We will only use one function, namely mesh that is available in base packages. Then we build the x and y coordinates, and create their Euclidean product with mesh. The variable ramp stores the values of the function $(x+y)/600$, which is already bounded to [0,1]. We then set the properties of an imagematrix object to these values, and plot them. Finally, the function imagematrixPNG produces a PNG file tailored to the size of its input image.

Figure 2.10 shows this ramp. Notice that $x \in \{1, \dots, 500\}$ denotes the lines, displayed from top to bottom, while the lines are stored in $y \in \{1, \dots, 100\}$ and displayed from left to right.

Some other convenient functions available in imagematrix are:

- clipping: maps any real value z into [0,1] by $\max\{0, \min\{z, 1\}\}$;
- normalize: maps all the values in the matrix $\boldsymbol{z}$ into [0,1] by $(z - \min(\boldsymbol{z}))/(\max(\boldsymbol{z}) - \min(\boldsymbol{z}))$;
- rgb2grey: converts a color image into gray levels;
- imagematrixEPS: produces an EPS file tailored to the size of its input image;
- equalize: equalizes all the values;
- equalize_indep: equalizes the values independently by the band;

Figure 2.10 Visualization of a ramp.

- `HistToEcdf`: computes an empirical cumulative distribution function from a histogram;
- `HistogramMatching`: transforms the destination image to have the same histogram as the reference image.

The reader is invited to browse through this source code for more information about these functions.

3

Intensity SAR Data and the Multiplicative Model

In this chapter, we derive the basic properties of Synthetic Aperture Radar (SAR) data, starting from the complex scattering vector and then reaching the Exponential and Gamma distributions. With this, we cover what many authors call *fully developed speckle*, or *speckle for textureless targets*. This will be generalized later for other situations.

A SAR sensor emits electromagnetic pulses and records the return from the target. It works in the microwaves region of the spectrum in different bands; see Table 3.1.

The bands are labeled with nonsequential letters and the bands are not regular. This was all done in the 2nd world war by the Allies to confuse the Axis. Efforts have been made to regularize the radar into decametric radar but they have all failed.

The frequency, and other parameters such as incidence angle, polarization, and imaging mode, influence the ability of the sensor to retrieve information from the target. As such, each band is used in a different kind of application:

X-band: Adequate for high-resolution imaging, as in mapping, agriculture, and ocean studies.
C-band: Penetrates tropical clouds and rain showers. Useful in sea ice surveillance and agriculture.
S-band: Rainfall measurement and airport surveillance.
L-band: Useful for agriculture, forestry, and soil moisture applications.
P-band: Unaffected by atmospheric effects, significant penetration through vegetation canopies, glacier or sea ice, and soil. Important for estimating vegetation biomass.

Assume that we illuminate an area with electromagnetic energy having N elementary backscatterers. Each backscatterer i will return a fraction of the incident energy A_i with phase ϕ_i. The total returned complex signal is, therefore,

$$S = \sum_{i=1}^{N} S_i = \sum_{i=1}^{N} A_i \exp\{\mathbf{j}\phi_i\} = \underbrace{\sum_{i=1}^{N} A_i \cos\phi_i}_{\Re(S)} + \mathbf{j}\underbrace{\sum_{i=1}^{N} A_i \sin\phi_i}_{\Im(S)}, \tag{3.1}$$

where $\mathbf{j} = \sqrt{-1}$ is the imaginary unit. This is usually called complex scattering. Figure 3.1 illustrates an example with $N = 15$ backscatterers (thin arrows) and the resulting complex scattering (thick arrow).

We have to make some assumptions in order to have a statistical description of the complex return S. These assumptions stem from the fact that we are using microwaves whose typical

SAR Image Analysis — A Computational Statistics Approach: With R Code, Data, and Applications, First Edition.
Alejandro C. Frery, Jie Wu, and Luis Gomez.

Companion website: www.wiley.com/go/frery/sarimageanalysis

Table 3.1 SAR Bands.

Band	Frequency	Wavelength
X-band	12.5 to 8 GHz	2.4 to 3.75 cm
C-band	8 to 4 GHz	3.75 to 7.5 cm
S-band	4 to 2 GHz	7.5 to 15 cm
L-band	2 to 1 GHz	15 to 30 cm
P-band	0.999 to 0.2998 GHz	30 to 100 cm

Source: Adapted from https://earth.esa.int/handbooks/asar/CNTR5-2.html.

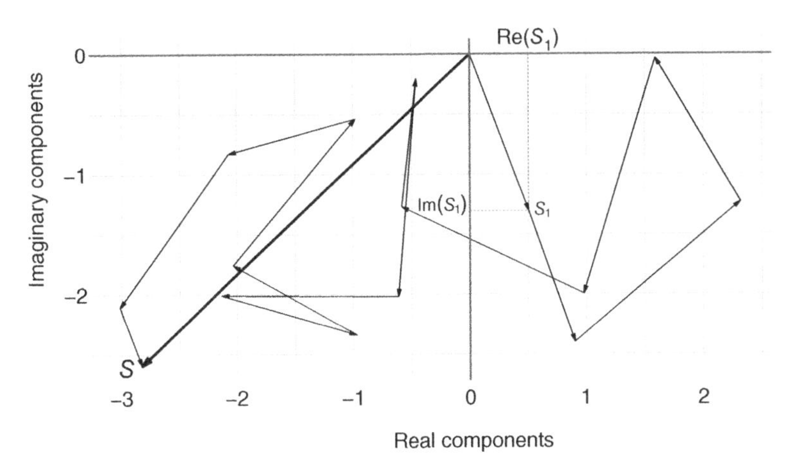

Figure 3.1 Example of complex scattering with $N = 15$ elementary backscatterers.

wavelength is of the order of centimeters, and spatial resolutions of the order decimeters. This leads to the

First assumption: We are observing backscatterers whose size is smaller than the wavelength, so $N \to \infty$; think, for instance, of the case of imaging a grass field.

Second assumption: There is no reason to expect that one or a few of these backscatterers dominate the return of the rest; there is no "mirror" in our resolution cell.

Third assumption: The (non-negative) amplitudes $A_i (i \in \{1,2,\dots,N\})$ are independent of each other.

Fourth assumption: There is no reason to expect any particular structure or phase dominance; with this, we may assume that the phases ϕ_i are outcomes of independent identically distributed Uniform random variables with support $(-\pi, \pi]$.

Fifth assumption: There is no association between phases and amplitudes.

With these hypotheses, it is possible to prove that the real and imaginary parts of S are both independent Gaussian random variables with zero mean and the same variance $\sigma^2/2$; σ^2 is often referred to as backscatter. Two targets with different backscatter only differ in the variance of their complex return S.

More often than not, instead of dealing directly with the complex scattering S one prefers to work with its amplitude $A = |S|$ or intensity $I = |S|^2$. Without loss of generality, we prefer the latter. If the real and imaginary parts of the complex scattering S are independent zero-mean

Gaussian random variables with variance $\sigma^2/2$, then the intensity follows an Exponential distribution with mean σ^2.

A unitary-mean exponential random variable has its distribution characterized by the density

$$f_Z(z) = e^{-z}\mathbb{1}_{(0,\mathbb{R}_+)}(z), \tag{3.2}$$

where $\mathbb{1}_A(z)$ is the indicator function of the set A, i.e. it takes value 1 inside A and zero otherwise. Denote this situation $Z \sim E(1)$ and notice that its expected value is $\mathrm{E}(Z) = 1$ and its variance is $\mathrm{Var}(Z) = 1$.

Being scale-invariant, if $Z' \sim E(1)$, then $Z = \sigma^2 Z'$ has density

$$f_Z(z) = \frac{1}{\sigma^2}e^{-z/\sigma^2}\mathbb{1}_{(0,\mathbb{R}_+)}(z), \tag{3.3}$$

and we denote this situation $Z \sim E(\sigma^2)$. The cumulative distribution function of $Z \sim E(\sigma^2)$ is

$$F_Z(z) = (1 - e^{-z/\sigma^2})\mathbb{1}_{(0,\mathbb{R}_+)}(z). \tag{3.4}$$

The mean and variance of $Z \sim E(\sigma^2)$ are, respectively, $\mathrm{E}(Z) = \sigma^2$ and $\mathrm{Var}(Z) = \sigma^4$. With these definitions, the coefficient of variation is one: $\mathrm{CV}(Z) = \sqrt{\mathrm{Var}(Z)}/\mathrm{E}(Z) = 1$.

In Figure 3.2 we show the densities, cumulative distribution functions and densities in semilogarithmic scale of three Exponential distributions, namely those with means equal to 1/2, 1, and 2.

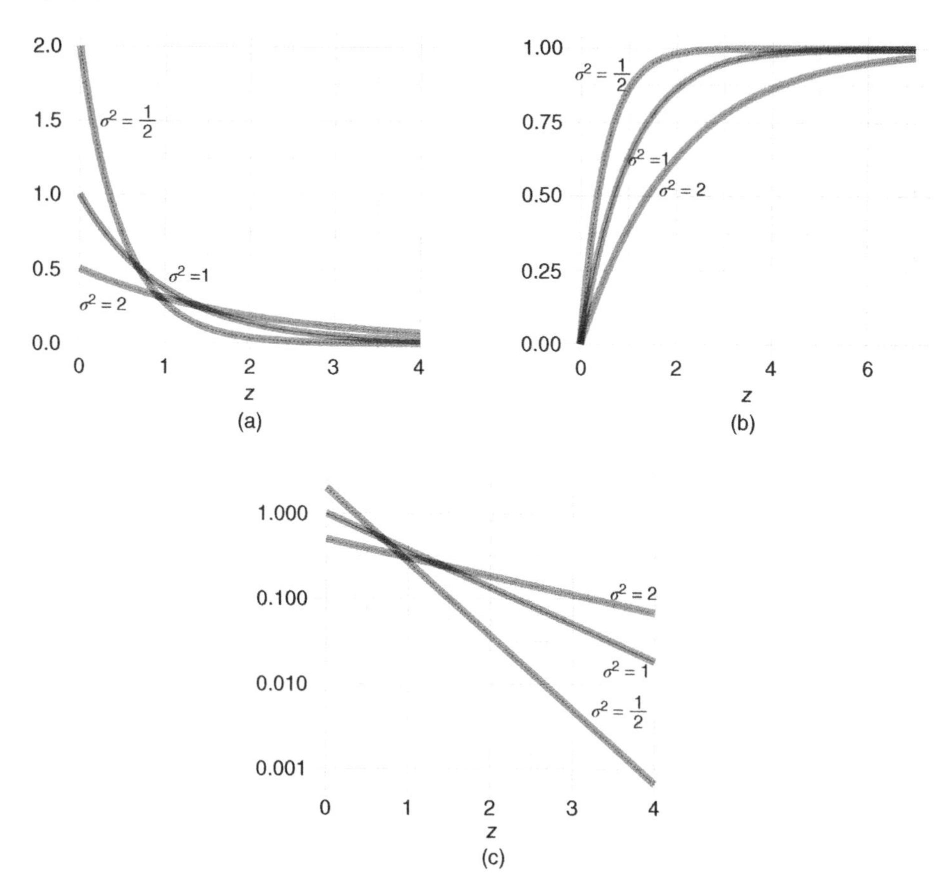

Figure 3.2 (a) Densities, (b) cumulative distribution functions, and (c) densities in semilogarithmic scale of the exponential distribution with means 1/2, 1, and 2 (dashed, solid, and dotted lines, resp.)

The semilogarithmic scale is particularly useful in revealing the tail behavior of distributions: this one is linear.

Multilook processing is often applied in order to improve the signal-to-noise ratio, which can be measured as the reciprocal of the coefficient of variation. It consists of using the mean of L independent observations:

$$Z = \frac{1}{L}\sum_{\ell=1}^{L} Z_\ell. \tag{3.5}$$

If each Z_ℓ follows an exponential distribution with mean σ^2, we have that Z obeys a Gamma distribution with mean σ^2 and shape parameter L. This distribution is characterized by the density

$$f_Z(z; L, \sigma^2) = \frac{L^L}{\sigma^{2L}\Gamma(L)} z^{L-1} \exp\left\{-Lz/\sigma^2\right\}, \tag{3.6}$$

where $\Gamma(v)$ is the Gamma function given by $\Gamma(v) = \int_{\mathbb{R}_+} t^{v-1}e^{-t}dt$. The reader is referred to Abramowitz and Stegun (1964) for details and relationships with other important special functions. We denote this situation as $Z \sim \Gamma(\sigma^2, L)$. This is also a scale-invariant distribution, in the sense that if $Z' \sim \Gamma(1, L)$, then $Z = \sigma^2 Z' \sim \Gamma(\sigma^2, L)$.

Figure 3.3 shows three cases of the Gamma distribution with unitary mean and shape parameters (Looks) equal to 1 (the Exponential distribution), 3, and 8.

Figure 3.3b is important, as this scale shows that the larger the number of looks, the less probable extreme events are. The Exponential density is linear and, thus, is a valuable reference when plotted along with other single-look intensity densities.

As we mentioned before, the intensity format is not the only possibility. Assume $Z \sim \Gamma(\sigma^2, L)$ and that $g : \mathbb{R}_+ \to \mathbb{R}$ is a monotonic function with inverse g^{-1}. The density of the random variable $W = g(Z)$ is given by

$$f_W(w; L, \sigma^2) = \frac{L^L}{\sigma^{2L}\Gamma(L)} \left|(g^{-1}(w))'\right| \left(g^{-1}(w)\right)^{L-1} \exp\left\{-Lg^{-1}(w)/\sigma^2\right\}. \tag{3.7}$$

Some authors Deledalle et al. (2009) prefer the amplitude format $W = \sqrt{Z}$, while other authors Santos et al. (2017) opt for a logarithmic transformation $W = \log(Z + 1)$ in order to

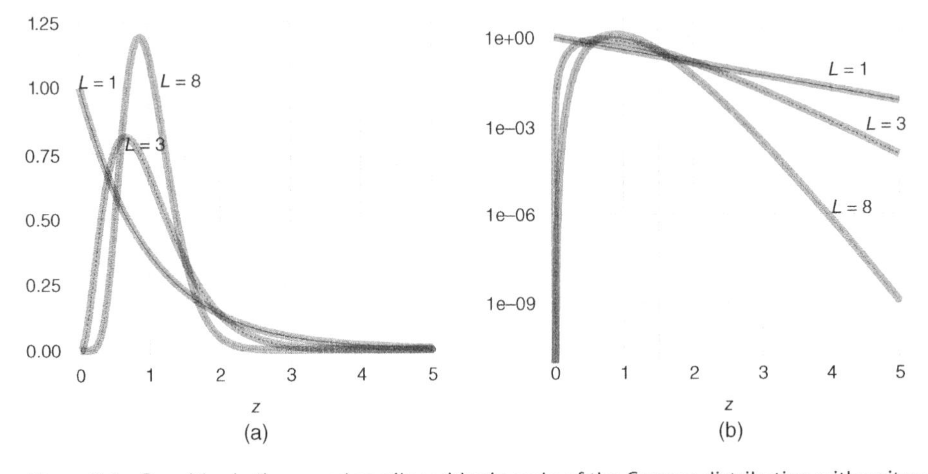

Figure 3.3 Densities in linear and semilogarithmic scale of the Gamma distribution with unitary mean and shape parameters 1, 3, and 8. (a) Densities and (b) densities in semilogarithmic scale.

Figure 3.4 Equalized first principal component of an ESAR image over Oberpfaffenhofen.

obtain an additive (although not Gaussian) model for the data. The model for the former is known as the *Square Root of Gamma* or *Nakagami* distribution, while the one for the latter is known as the *Fisher-Tippet* distribution when $L = 1$. The reader is invited to obtain these densities using (3.7).

In the following we illustrate the adequacy of this model with data from an actual sensor.

Figure 3.4 shows an enhanced version of the first principal component obtained from three amplitude bands (each corresponding to one polarization) on the surroundings of Oberpfaffenhofen by the Experimental Synthetic Aperture Radar (ESAR) sensor.

The image shown in Figure 3.4 has 1599×4000 pixels. We selected the 15 561 positions within the black rectangle in the upper part of the image as representative of a textureless region, and extracted the observations both in the complex and intensity channels (without equalization). This area, which is delimited with a clear rectangle, was chosen because there is no visual clue of texture in it.

Figure 3.5 shows the histograms and fitted Gaussian densities of the six data sets corresponding to the real and imaginary parts of each polarization. Notice that there is no evidence of either skewness or lack/excess of dispersion with respect to the hypothesized model. For the sake of visualization, these histograms were produced with a fraction of the number of bins prescribed by the Freedman-Diaconis rule (Freedman and Diaconis, 1981).

Figure 3.6 shows the histograms and fitted Exponential densities of these three data sets.

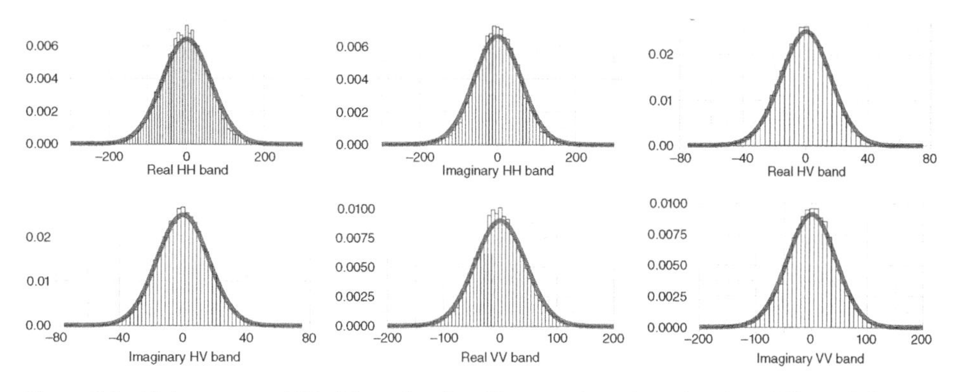

Figure 3.5 Histograms and fitted Gaussian densities over the selected region of Figure 3.4.

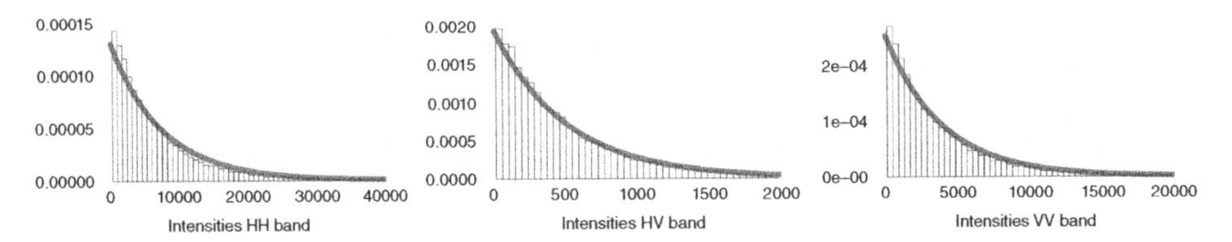

Figure 3.6 Histograms and fitted Exponential densities over the dark region of Figure 3.4.

The quality of the fit is remarkable, in particular for such large number of observations. This is an empirical indication that the model we are considering is adequate.

One of the most famous statistical aphorisms is due to Prof. George Box:

> *"The most that can be expected from any model is that it can supply a useful approximation to reality:*
> ***All models are wrong; some models are useful."***
>
> Box et al. (2005)

In line with this idea, in this chapter we will discuss the most important and illuminating models for SAR data that generalize the one for textureless data.

The Multiplicative Model is just one of the infinitely many ways to build stochastic descriptions for SAR data. Among its advantages we would like to mention that it can be considered an *ab initio* model, and that it leads to expressive and tractable descriptions of the data.

Let us recall that the basic model for multilook intensity data is the $\Gamma(\sigma^2, L)$ law whose density is

$$f_Z(z; L, \sigma^2) = \frac{L^L}{\sigma^{2L}\Gamma(L)} z^{L-1} \exp\left\{-Lz/\sigma^2\right\}. \tag{3.8}$$

As previously said, the Gamma distribution is scale-invariant, so we may pose this model as the product between the constant backscatter $X = \sigma^2$ and the multilook speckle $Y \sim \Gamma(1, L)$.

But, are there situations where we cannot assume a constant backscatter? Yes, there are.

A constant backscatter results from infinitely many elementary backscatterers, i.e. from the assumption that $N \to \infty$ in (3.1). Such an assumption makes the particular choice of the sensed area irrelevant. But this may not be the case always.

The advent of higher resolution sensors makes this hypothesis unsuitable in areas where the elementary backscatterers are of the order of the wavelength; cf. Table 3.1. If, for instance, we are dealing with a 1 m × 1 m resolution image, we may consider $N \to \infty$ if the target is flat and composed of grass; but if the target is a forest, this assumption may be unrealistic.

3.1 The $\mathcal{K}$ Distribution

Jakeman and Pusey (1976) were among the first who tackled this problem. Assuming that the number of elementary backscatterers N fluctuates according to a Negative Binomial distribution (see Section 4.1.3), they obtained a closed-form density which characterizes the $\mathcal{K}$ distribution:

$$f_Z(z; \alpha, \lambda, L) = \frac{2\lambda L}{\Gamma(\alpha)\Gamma(L)} (\lambda L z)^{\frac{\alpha+L}{2}-1} K_{\alpha-L}(2\sqrt{\lambda L z}), \tag{3.9}$$

where $\alpha > 0$ measures the roughness, $\lambda > 0$ is a scale parameter, and K_ν is the modified Bessel function of order ν. This special function is given by $K_\nu(z) = \int_0^\infty e^{-z}\cosh(\nu t)dt$. See the book by Gradshteyn and Ryzhik (1980) for other definitions and important properties. This function is implemented in many numerical platforms as, for instance, in R.

We denote $Z \sim \mathcal{K}(\alpha, \lambda, L)$ the situation of Z following the distribution characterized by (3.9). The k-order moments of Z are

$$E(Z^k) = (\lambda L)^{-k}\frac{\Gamma(L+k)\Gamma(\alpha+k)}{\Gamma(L)\Gamma(\alpha)}. \tag{3.10}$$

Eq. (3.10) is useful, among other applications, for finding $\lambda^* = \alpha$, the scale parameter that yields a unitary mean distribution for each α and any L.

Figure 3.7 shows the densities in linear and semilogarithmic scales of the Exponential and $\mathcal{K}$ distributions. They all have unitary mean, and the latter is shown with different degrees of roughness ($\alpha \in \{1.5, 3, 10\}$). It is noticeable that the larger the value of α is, the closer the $\mathcal{K}$ and E densities become. In fact, Frery et al. (1997) prove that there is convergence in distribution of the latter to the former.

The difference between these distributions is noteworthy, c.f. Figure 3.7b. The solid straight line is the density of the Exponential distribution, while the dash-dot line is that of the $\mathcal{K}(1.5, 1.5, 1)$ law. The latter assigns larger probabilities to both small and large values, when compared with the former. This leads, as will be seen later, to very contrasted data.

Figure 3.8 shows the effect of varying the number of looks, for the same $\alpha = 2$ and $\lambda = 2$. Notice in Figure 3.8b the dramatic effect multilook processing has mostly on the distribution of very small values. This, along with the reduced probability very large values have with multilook processing, yields less contrasted images.

Although the basic construction, and physical explanation, of the $\mathcal{K}$ distribution stems from letting the number of elementary backscatterers fluctuate, we are interested in an equivalent derivation.

As said previously, the basic model for observations without roughness is (4.9). The mean $X = \sigma^2$ can be seen as multiplying Y, a $\Gamma(1, L)$ random variable, which describes the speckle. As we are interested in letting the mean fluctuate, X can be described by any distribution

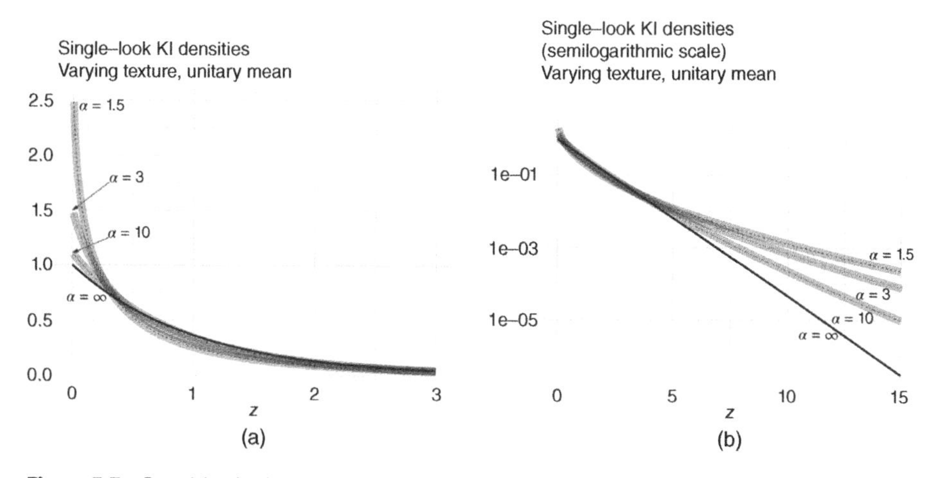

Figure 3.7 Densities in linear and semilogarithmic scales of the E(1) (solid) and $\mathcal{K}$ distributions with unitary mean ($\alpha \in \{1.5, 3, 10\}$ in dashed, dotted, and dash-dot lines, resp. (a) Densities and (b) densities in semilogarithmic scale.)

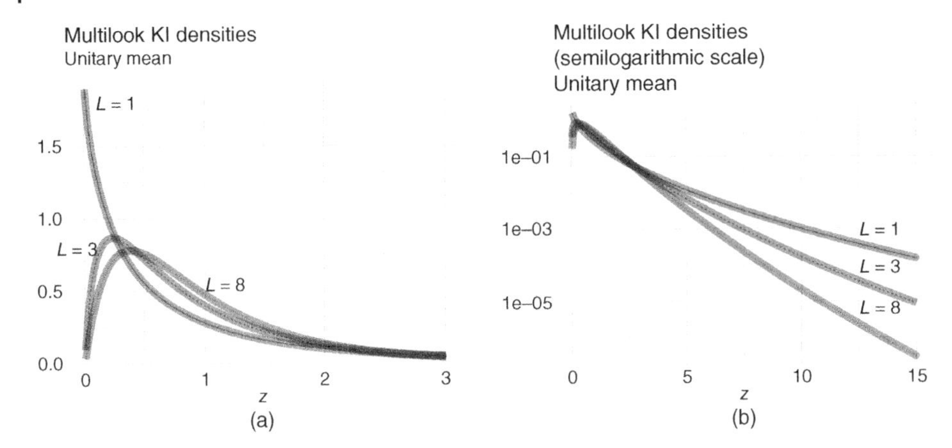

Figure 3.8 Densities in linear and semilogarithmic scale $\mathcal{K}(2,2,L)$ distributions with unitary mean and $L \in \{1,3,8\}$ in solid, dashed, and dotted line, resp. (a) Densities and (b) densities in semilogarithmic scale.

with positive support. If we choose a Gamma random variable with mean $\alpha/\lambda > 0$ and shape parameter $\alpha > 0$, i.e. $Y \sim \Gamma(\alpha/\lambda, \alpha)$, and further assume that X and Y are independent, then the product $Z = XY$ follows a $\mathcal{K}(\alpha, \lambda, L)$ distribution with density (3.9). This *multiplicative* construction is not only useful for sampling from this distribution, but also for obtaining other models for the return.

3.2 The $\mathcal{G}^0$ Distribution

Frery et al. (1997) noticed that the $\mathcal{K}$ distribution failed to describe data from extremely textured areas as, for instance, urban targets. The authors then proposed a different model for the backscatter X: the Reciprocal Gamma distribution.

We say that $X \sim \Gamma^{-1}(\alpha, \gamma)$, with $\alpha < 0$ and $\gamma > 0$ follows a Reciprocal Gamma distribution if its density is

$$f_X(x; \alpha, \gamma) = \frac{\gamma^{-\alpha}}{\Gamma(-\alpha)} x^{\alpha-1} \exp\{-\gamma/x\}, \tag{3.11}$$

for $x > 0$ and zero otherwise.

Now introducing the Reciprocal Gamma model for the backscatter in the multiplicative model, i.e. by multiplying the independent random variables $X \sim \Gamma^{-1}(\alpha, \gamma)$ and $Y \sim \Gamma(1, L)$, one obtains the $\mathcal{G}^0$ distribution for the return $Z = XY$, which is characterized by the density

$$f_Z(z; \alpha, \gamma, L) = \frac{L^L \Gamma(L-\alpha)}{\gamma^\alpha \Gamma(L)\Gamma(-\alpha)} \frac{z^{L-1}}{(\gamma + Lz)^{L-\alpha}}, \tag{3.12}$$

where $\alpha < 0$, and $\gamma, z > 0$. It is noteworthy that, differently from (3.9), this density does not involve Bessel functions.

We denote $Z \sim \mathcal{G}^0(\alpha, \gamma, L)$ the situation of Z following the distribution characterized by (3.12). The k-order moments of Z are

$$E(Z^k) = (\gamma/L)^k \frac{\Gamma(L+k)\Gamma(-\alpha-k)}{\Gamma(L)\Gamma(-\alpha)}, \tag{3.13}$$

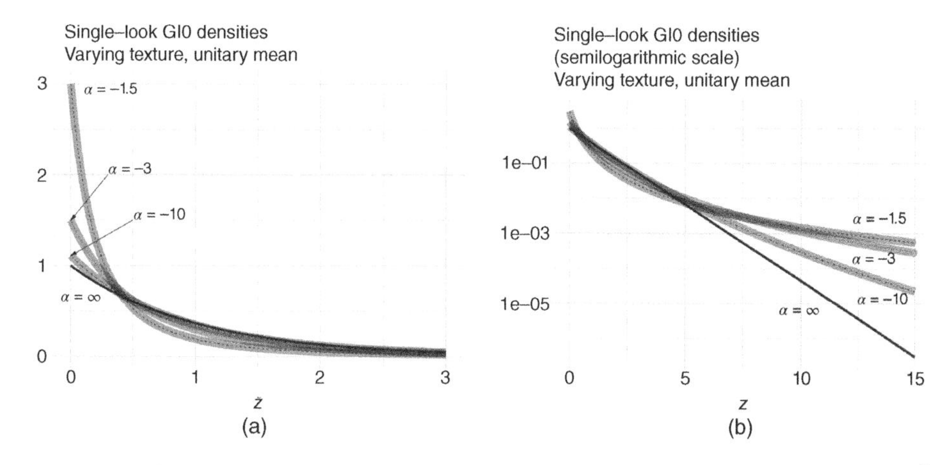

Figure 3.9 Densities in linear and semi-logarithmic scale of the E(1) (solid) and single-look $\mathcal{G}^0$ distributions with unitary mean and $\alpha \in \{-1.5, -3, -10\}$ in dashed, dotted, and dash-dot lines, resp. (a) Densities and (b) densities in semilogarithmic scale.

provided $-\alpha > k$, and infinite otherwise. Eq. (3.13) is useful, among other applications, for finding $\gamma^* = -\alpha - 1$, the scale parameter that yields a unitary mean distribution for each α and any L.

Figure 3.9 shows the E(1) and $\mathcal{G}^0(\alpha, \gamma^*, 1)$ densities, with $\alpha \in \{-1.5, -3, -10\}$. The differences in tail behavior are clearly exhibited in the semilogarithmic scale; cf. Figure 3.9b. Whereas the exponential distribution decreases linearly, the $\mathcal{G}^0$ law assigns more probability to larger events increasing the variability of the return.

Figure 3.10 shows the effect of varying the number of looks, for the same $\alpha = -5$ and $\gamma = 4$. Notice, again, in Figure 3.10b the effect multilook processing has mostly on the distribution of very small values. This, along with the reduced probability very large values have with multilook processing, yields less contrasted images.

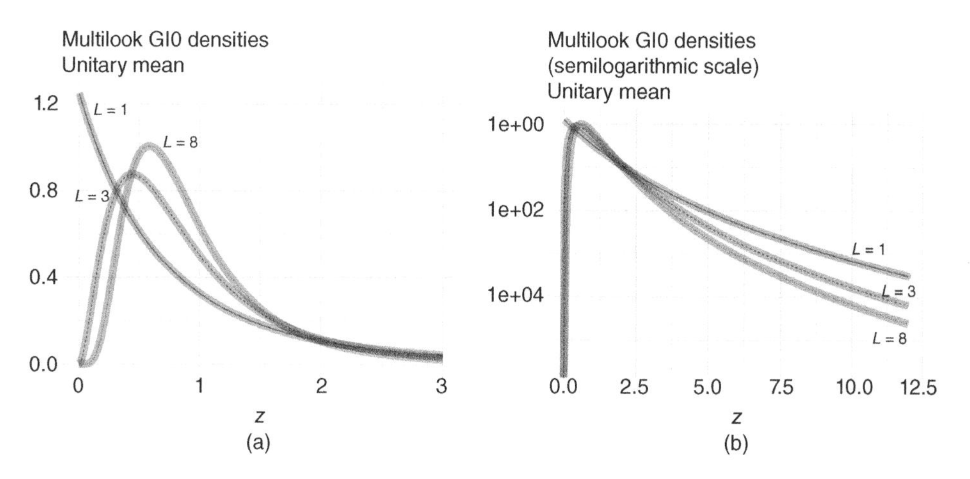

Figure 3.10 Densities in linear and semilogarithmic scale $\mathcal{G}^0(-5, 4, L)$ distributions with unitary mean and $L \in \{1, 3, 8\}$ in solid, dashed, and dotted lines, resp. (a) Densities and (b) densities in semilogarithmic scale.)

The $\mathcal{G}^0$ distribution has the same number of parameters as the $\mathcal{K}$ law, but it has been shown to be more apt at modeling data with extreme variability. Moreover, it is also able to describe the same kind of images from textured areas for which the latter was proposed; Mejail et al. (2003) showed that, with proper choices of parameters, the $\mathcal{G}^0$ law can approximate with any error any $\mathcal{K}$ distribution (see details in Mejail et al., 2001). For these reasons, the $\mathcal{G}^0$ distribution is called *Universal Model* for SAR data; see Exercise 13.

The $\mathcal{G}^0$ distribution relates to the well-known Fisher-Snedekor law in the following manner:

$$F_{G^0(\alpha,\gamma,L)}(t) = \Upsilon_{2L,-2\alpha}(-\alpha t/\gamma), \tag{3.14}$$

where $\Upsilon_{u,v}$ is the cumulative distribution function of a Fisher-Snedekor distribution with u and v degrees of freedom, and $F_{G^0(\alpha,\gamma,L)}$ is the cumulative distribution function of a $F_{G^0(\alpha,\gamma,L)}$ random variable. Notice that Υ is readily available in most software platforms for statistical computing. Since such platforms usually also provide implementations of the inverse of cumulative distribution functions, the Inversion Theorem can be used to sample from the $\mathcal{G}^0$ law.

Arguably, the most popular way of sampling from the $\mathcal{G}^0$ distribution is through its multiplicative nature. Obtaining deviates from $X \sim \Gamma(1, L)$ is immediate. In order to sample from $Y \sim \Gamma^{-1}(\alpha, \gamma)$, one may use the fact that if $Y' \sim \Gamma(-\alpha, \gamma)$, then $Y = 1/Y'$ has $\Gamma^{-1}(\alpha, \gamma)$ distribution. Then, $Z = X/Y'$ has the desired $\mathcal{G}^0(\alpha, \gamma, L)$ distribution.

Chan et al. (2018) discuss other techniques for obtaining such deviates, in particular interesting connections between the $\mathcal{G}^0(\alpha, \gamma, 1)$ law and certain Pareto distribution.

3.3 The $\mathcal{G}^H$ Distribution

Frery et al. (2010) used the Inverse Gaussian distribution to describe the backscatter. We say that $X \sim \text{IG}(\lambda, \mu)$, with $\lambda, \mu > 0$, follows such a model if its density is

$$f_X(x; \lambda, \mu) = \sqrt{\frac{\lambda}{2\pi x^3}} \exp\left\{-\frac{\left(\sqrt{\mu x} - \sqrt{\lambda}\right)^2}{2x}\right\}, \tag{3.15}$$

for $x > 0$ and zero otherwise. If $X \sim \text{IG}(\lambda, \mu)$ and $Y \sim \Gamma(1, L)$ are independent random variables, then the intensity return follows a $\mathcal{G}^H$ distribution, characterized by the density

$$f_Z(z; \lambda, \mu, L) = \frac{L^L e^{\lambda}}{\Gamma(L)} \sqrt{\frac{2\lambda\mu}{\pi}} \left[\frac{\lambda}{(\mu\lambda + 2Lz)\mu}\right]^{\frac{2L+1}{4}} z^{L-1} K_{L+\frac{1}{2}} \sqrt{\frac{\lambda}{\mu}(\lambda + 2Lz)}, \tag{3.16}$$

if $z > 0$ and zero otherwise. We denote this situation $Z \sim \mathcal{G}^H(\lambda, \mu, L)$.

The advantage of (3.16) over (3.9) is that the Bessel function can be computed without relying on its numerical implementation. If n is an integer, then

$$K_{n+\frac{1}{2}}(v) = e^v \sqrt{\frac{\pi}{2v}} \sum_{k=0}^{n} \frac{(n+k)!}{k!(n-k)(2v)^k}.$$

The k-order moment of $Z \sim \mathcal{G}^H(\lambda, \mu, L)$ is

$$E(Z^k) = \sqrt{\frac{2\lambda}{\pi}} \left(\frac{\sqrt{\mu}}{L}\right)^k \exp\left\{\frac{\lambda}{\sqrt{\mu}}\right\} \mu^{-\frac{1}{4}} \frac{\Gamma(L+k)}{\Gamma(L)} K_{k-\frac{1}{2}}\left(\frac{\lambda}{\sqrt{\mu}}\right). \tag{3.17}$$

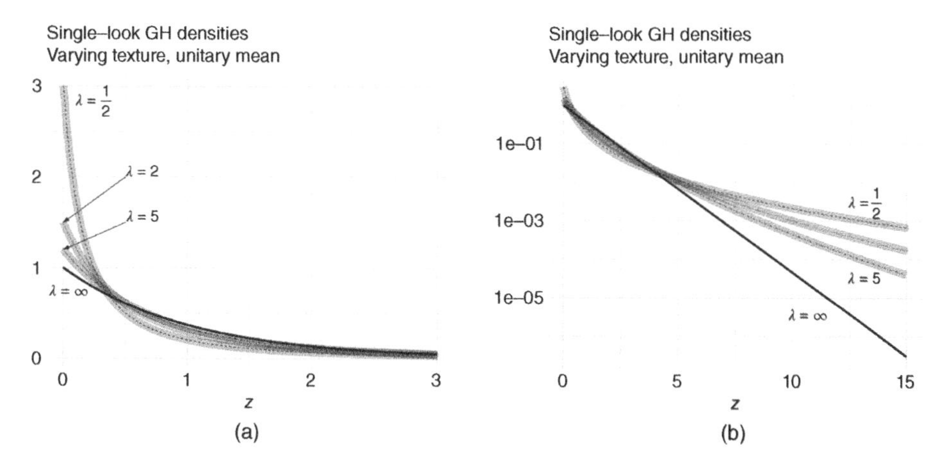

Figure 3.11 Densities in linear and semi-logarithmic scale of the E(1) (solid) and single-look $\mathcal{G}^H$ distributions with unitary mean and $\omega \in \{1/2, 2, 5\}$ in dashed, dotted, and dash-dot lines, resp. (a) Densities and (b) densities in semilogarithmic scale.

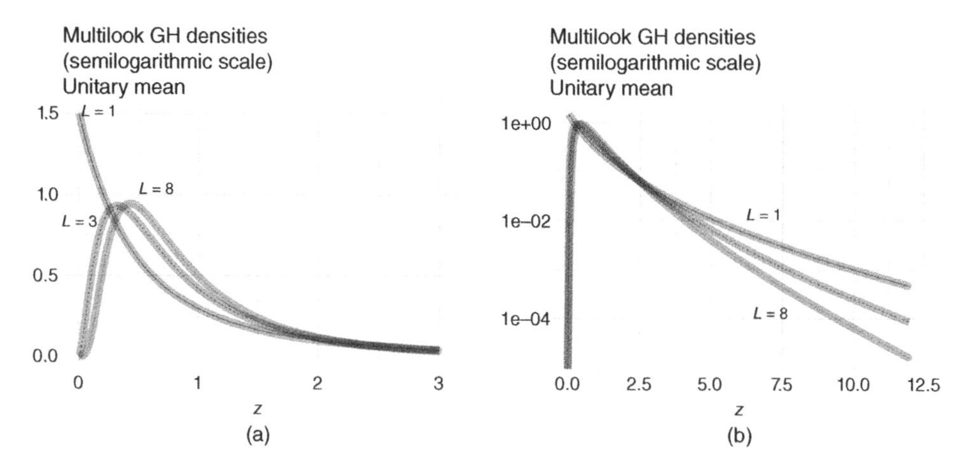

Figure 3.12 Densities in linear and semi-logarithmic scale of $\mathcal{G}^H(2, 1, L)$ distributions with $L \in \{1, 3, 8\}$ in solid, dashed, and dotted lines, resp. (a) Densities and (b) densities in semilogarithmic scale.

Notice that $E(Z) = \mu$.

Figure 3.11 shows the E(1) and $\mathcal{G}^H(\lambda, 1, 1)$ densities, with $\lambda \in \{1/2, 2, 5\}$. The differences in tail behavior are clearly exhibited in the semilogarithmic scale; cf. Figure 3.11b. Whereas the exponential distribution decreases linearly, the $\mathcal{G}^H$ law assigns more probability to larger events increasing the variability of the return.

Figure 3.12 shows that multilooking has a similar effect on the $\mathcal{G}^H$ distribution to that already observed in the $\mathcal{G}^0$ and $\mathcal{K}$ laws.

3.4 Connection Between Models

The Γ, $\mathcal{K}$, $\mathcal{G}^0$, and $\mathcal{G}^H$ distributions are particular cases of a more general law: the $\mathcal{G}$ distribution. This last one is characterized by four parameters: α, λ, γ, and the number of looks

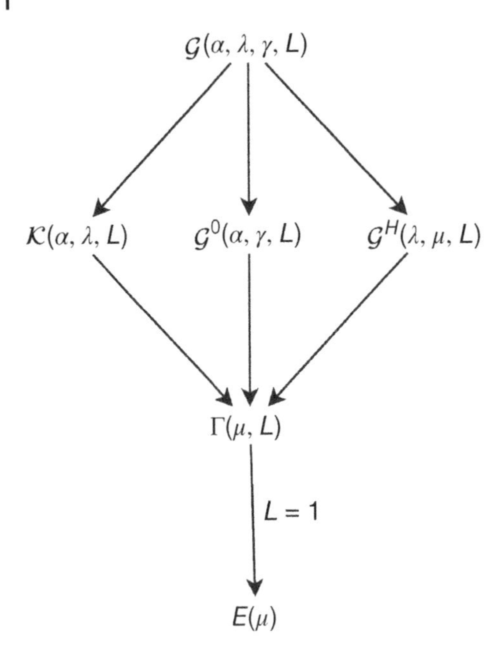

Figure 3.13 Relationships between the $\mathcal{G}$, $\mathcal{K}$, $\mathcal{G}^0$, $\mathcal{G}^H$, Γ, and Exponential distributions.

$L \geq 1$ (Frery et al., 1997). The difficulty in estimating them has prevented its use in practical situations. Figure 3.13 illustrates how these distributions relate.

Apart from this kind of relationship, where a distribution is a particular case of another, Delignon and Pieczynski (2002) provide a different approach.

The authors start from Eq. (3.1), the equation that describes the complex return, and relax the first assumption. Instead of assuming that $N \to \infty$, they consider that N follows a Poisson distribution with mean η. This mean is then allowed to vary according to four probability laws, giving rise to the Γ, $\mathcal{K}$, $\mathcal{G}^0$, KU, and W distributions for the return.

This last approach is the most general available, as it also allows introducing several types of correlation structures. The readers are referred to the works by Yue et al. (2020), Yue et al. (2021b), and Yue et al. (2021a) for details.

Exercises

1 Find the densities of the Nakagami and Fisher-Tippet distributions. Illustrate them.

2 Assume Z follows a Nakagami distribution. Find the density of $W = \log(Z + 1)$ (a multilook Fisher-Tippet distribution). Illustrate.

3 Using (3.10), compute the skewness and kurtosis of the $\mathcal{K}(\alpha, \lambda, L)$ distribution. Illustrate how they vary with the parameters.

4 Illustrate the dependence of the density of the $\mathcal{K}$ distribution with respect to the scale parameter λ.

5 Obtain samples from iid $\mathcal{K}(\alpha, \lambda, L)$ random variables. Produce histograms, and draw the theoretical densities over them for a variety of parameters.

6 Obtain the expression of the density of $\mathcal{K}$-distributed amplitude data via the transformation $Z_A = \sqrt{Z_I}$. Compute its moments. Illustrate.

7 Using (3.13), compute the skewness and kurtosis of the $\mathcal{G}^0(\alpha, \gamma, L)$ distribution. Illustrate how they vary with the parameters.

8 Illustrate the dependence of the density of the $\mathcal{G}^0$ distribution with respect to the scale parameter γ.

9 Obtain samples from iid $\mathcal{G}^0(\alpha, \gamma, L)$ random variables. Produce histograms, and draw the theoretical densities over them for a variety of parameters.

10 Obtain the expression of the density of $\mathcal{G}^0$-distributed amplitude data via the transformation $Z_A = \sqrt{Z_I}$. Compute its moments. Illustrate.

11 Consider (3.9) and (3.10). Compute μ, the expected value, from the latter, and reparametrize the former using μ, α, and L.

12 Consider (3.12) and (3.13). Compute μ, the expected value, from the latter, and reparametrize the former using μ, α, and L.

13 Propose a measure of the difference between $Z_1 \sim \mathcal{K}(\alpha_1, \lambda^*, L)$ and $Z_2 \sim \mathcal{G}^0(\alpha_2, \gamma^*, L)$ (consider, for instance, the Hellinger distance between distributions with positive support $d_{\text{H}}(Z_1, Z_2) = 1 - \int_{\mathbb{R}_+} \sqrt{f_{Z_1} f_{Z_2}}$). Describe this difference for a number of values of α_1 and α_2, for $L \in \{1,3,8\}$. For each α_1, find the value of α_2 which yields the closest $\mathcal{K}$ and $\mathcal{G}^0$ random variables (Mejail et al., 2001).

4

Parameter Estimation

At this point we have data and models. We will see ways of using the former to make inferences about the latter.

Assume we have a parametric model $(\Omega, \mathcal{A}, \Pr)$ for real independent random variables X and $\boldsymbol{X} = (X_1, X_2, \dots)$. The distribution of X is indexed by $\boldsymbol{\theta} \in \boldsymbol{\Theta} \subset \mathbb{R}^p$, where $p \geq 1$ is the dimension of the parametric space $\boldsymbol{\Theta}$.

What can we say about $\boldsymbol{\theta}$ using the sample $\boldsymbol{X}$? This is parametric statistical inference!

We will later need an additional ingredient: expected values of transformations of the random variable X:

$$\mathrm{E}\boldsymbol{\psi}(X) = \left(\mathrm{E}\psi_1(X), \mathrm{E}\psi_2(X), \dots, \mathrm{E}\psi_p(X)\right), \tag{4.1}$$

where each ψ_j is a measurable function $\psi_j : \mathbb{R} \to \mathbb{R}$. Each element of (4.1) is given by

$$\mathrm{E}\psi_j(X) = \int_R \psi_j(x) dF(x), \tag{4.2}$$

and F is the cumulative distribution function of X. If $\psi(X) = X^k$, we say that $\mathrm{E}X^k$ is the k-th order moment of X (if it exists).

The quantity $\mathrm{E}(X - \mathrm{E}X)^k$ is called "the central moment" of X, if it exists. The second central moment $\mathrm{E}(X - \mathrm{E}X)^2 = \mathrm{E}X^2 - (\mathrm{E}X)^2$ is called "the variance" of X. We denote it $\mathrm{Var}X$.

In general, $\mathrm{E}X^k \neq (\mathrm{E}X)^k$.

4.1 Models

We will use a few models as examples. The interested reader is referred to, among other references, to the books by Johnson et al. (1993) for discrete distributions, and by Johnson et al. (1995) for continuous random variables.

4.1.1 The Bernoulli Distribution

This is the basic discrete distribution. It models the outcome of a dichotomic random experiment, i.e. with only two possible outcomes: failure (0) or success (1). Denote $N \sim \mathrm{Be}(p)$ a random variable with Bernoulli distribution and probability $0 \leq p \leq 1$. The vector of probabilities is $(1 - p, p)$.

SAR Image Analysis — A Computational Statistics Approach: With R Code, Data, and Applications, First Edition.
Alejandro C. Frery, Jie Wu, and Luis Gomez.

Companion website: www.wiley.com/go/frery/sarimageanalysis

4.1.2 The Binomial Distribution

A binomial random variable is the result of counting the successes in n independent identically distributed Bernoulli trials.

If $X \sim \text{Bi}(p, n)$, its distribution is characterized by the probability function:

$$\Pr(X = k) = \binom{n}{k} p^k (1-p)^{n-k},$$

in which $k \in \{0, 1, \ldots, n\}$, and $0 < p < 1$.

4.1.3 The Negative Binomial Distribution

This discrete distribution is basic in the construction of the $\mathcal{K}$ law. We say that $N \sim \text{NB}(k, p)$ has a negative binomial distribution if its vector of probabilities is given by

$$\Pr(N = n) = \binom{n-1}{k-1} p^k (1-p)^{n-k},$$

with $0 < p < 1$, $n \geq k$, and $k \in \mathbb{N}_0$ (the natural numbers including zero).

This is the distribution that models the probability of having to wait until the n-th Bernoulli trial with probability p of success until we observe exactly k failures.

4.1.4 The Uniform Distribution

We say that $X \sim \mathcal{U}_{(0,\theta)}$, $\theta > 0$ has uniform distribution over the interval $(0, \theta)$ when its density is

$$f_U(u; \theta) = \frac{1}{\theta} \mathbb{1}_{(0,\theta)}(u). \tag{4.3}$$

With this, we have that the k-order moment of X is

$$\text{E}U^k = \int_{\mathbb{R}} \frac{1}{\theta} \mathbb{1}_{(0,\theta)} u^k d(u) du = \int_0^{\theta} \frac{1}{\theta} u^k du = \frac{1}{k+1} \theta^k. \tag{4.4}$$

Its cumulative distribution function is

$$F_U(u; \theta) = \begin{cases} 0 & \text{if } u \leq 0, \\ u/\theta & \text{if } 0 < u < \theta, \\ 1 & \text{otherwise.} \end{cases} \tag{4.5}$$

4.1.5 Beta Distribution

Closely related to the Uniform distribution is the Beta law (Devroye, 1986). We say that B follows a Beta distribution with shape parameters $a, b > 0$ if its density is given by

$$f_B(u; a, b) = \frac{\Gamma(a+b)}{\Gamma(a)\Gamma(b)} u^{a-1} (1-u)^{b-1} \mathbb{1}_{(0,1)}(u). \tag{4.6}$$

We denote this situation as $B \sim \mathcal{B}(a, b)$. Its expected value and variance are given by:

$$\text{E}(B) = \frac{a}{a+b} \text{ and } \text{Var}(B) = \frac{ab}{(a+b)^2(a+b+1)}.$$

It is noteworthy that $\mathcal{B}(1, 1)$ is the Uniform distribution, and that if $a = b$ the density is symmetric around $1/2$. If $a = b > 1$ the unique mode is at $1/2$, otherwise there are two modes at

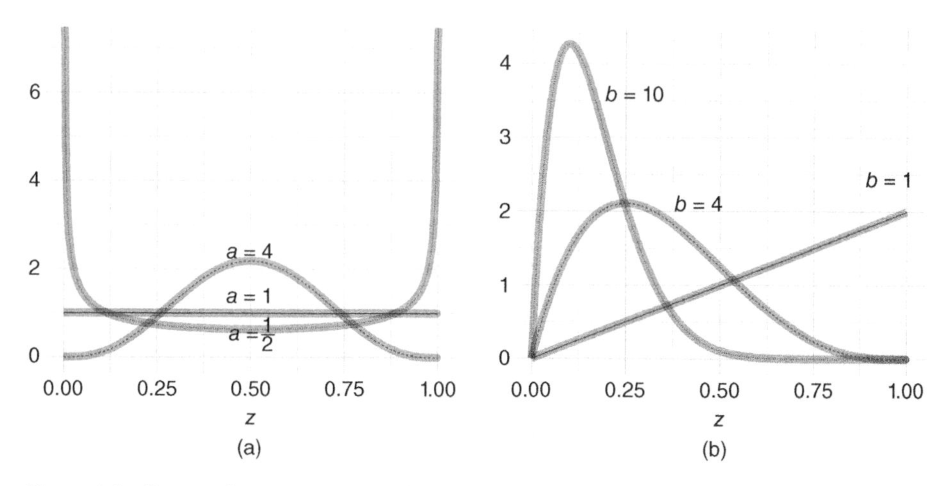

Figure 4.1 Symmetric and asymmetric Beta densities. (a) Symmetric Beta densities ($a = b \in \{0.5, 1, 4\}$). (b) Asymmetric Beta densities ($a = 2, b \in \{1, 4, 10\}$).

0 and 1. Figure 4.1a illustrates three of these situations. Figure 4.1b shows three Beta densities with $a = 2$, and $b \in \{1, 4, 10\}$.

Two properties make the Beta distribution specially apt for image processing operations: its finite support, and its flexibility. With this, as presented in Chapter 1, we may use the Inversion Theorem and stipulate the a histogram with properties as desired.

4.1.6 The Gaussian Distribution

We say that $X \sim N(\mu, \sigma^2)$ follows a Gaussian distribution with mean $\mu \in \mathbb{R}$ and variance σ^2 if the density that characterizes its distribution is

$$f_G(x; \mu, \sigma^2) = \frac{1}{\sqrt{2\pi}\sigma} \exp\left\{-\frac{1}{2\sigma^2}(x-\mu)^2\right\}. \tag{4.7}$$

Its first and second moment are $\mathrm{E}(X) = \mu$ and $\mathrm{E}(X^2) = \mu^2 + \sigma^2$.

We don't know explicit expressions for its cumulative distribution function, but it is widely implemented in almost every statistical platform.

4.1.7 Mixture of Gaussian Distributions

A mixture of n Gaussian models is characterized by the density

$$f_{\mathrm{MG}}\left(x; \boldsymbol{p}, \boldsymbol{\mu}, \boldsymbol{\sigma}^2\right) = \sum_{j=1}^{n} \frac{p_j}{\sqrt{2\pi}\sigma_j} \exp\left\{-\frac{1}{2\sigma_j^2}\left(x-\mu_j\right)^2\right\}, \tag{4.8}$$

where $\boldsymbol{p} = \left(p_1, p_2, \dots, p_n\right)$ is the vector of probabilities, $\boldsymbol{\mu} = \left(\mu_1, \mu_2, \dots, \mu_n\right)$ is the vector of means, and $\boldsymbol{\sigma}^2 = \left(\sigma_1^2, \sigma_2^2, \dots, \sigma_n^2\right)$ is the vector of variances. In this way, the parameter space is $\boldsymbol{\Theta} = S^n \times R^n \times R^n_+$, where S^n is the surface of the n-dimensional simplex. This space is a subset of $\mathbb{R}^{(n-1)n^2}$. We denote this situation $X \sim \mathrm{MG}\left(\boldsymbol{p}, \boldsymbol{\mu}, \boldsymbol{\sigma}^2\right)$.

Using the fact that the random components are independent, it is immediate that $\mathrm{E}X = \sum_{j=1}^{n} p_j\mu_j$ and that $\mathrm{Var}\, X = \sum_{j=1}^{n} p_j\sigma_j^2$.

4.1.8 The (SAR) Gamma Distribution

The Gamma distribution, in the usual parametrization for Synthetic Aperture Radar (SAR) data $Z \sim \Gamma(\mu, L)$, with $L, \mu > 0$ is characterized by the density

$$f_\Gamma(z; L, \mu) = \frac{L^L}{\mu^L \Gamma(L)} z^{L-1} \exp\{-Lz/\mu\}. \tag{4.9}$$

With this, the first- and second-order moments of Z are $\mathrm{E}Z = \mu$, and $\mathrm{Var}Z = \mu^2/L$, respectively.

As for the Gaussian distribution, in general, we do not have explicit expressions for its cumulative distribution function, with the exponential case ($L = 1$) an exception.

This is a good model for the intensity of SAR observations over areas with no texture. The multiplicative model says the observed return Z is the product of two independent random variables: X, the backscatter, and Y, the speckle. The speckle can be modeled by a Gamma random variable with shape parameter $L \geq 1$ (the number of looks) and unitary mean.

If the area under observation has constant backscatter ($X = \mu$), then the return $Z \sim \Gamma(\mu, L)$.

4.1.9 The Reciprocal Gamma Distribution

When the backscatter varies in the observed area we say that the area has texture. A very flexible model for the backscatter is the Reciprocal Gamma law (Frery et al., 1997), which is characterized by the density

$$f_Y(y; \alpha, \gamma) = \frac{1}{\gamma^\alpha \Gamma(-\alpha)} y^{\alpha-1} \exp\left\{-\frac{\gamma}{y}\right\}, \tag{4.10}$$

where $\alpha < 0$ is the texture, and $\gamma > 0$ is the scale. We denote this situation $Y \sim \Gamma^{-1}(\alpha, \gamma)$.

4.1.10 The $\mathcal{G}_I^0$ Distribution

Assuming that the backscatter is $X \sim \Gamma^{-1}(\alpha, \gamma)$, the speckle is $Y \sim \Gamma(1, L)$, and that they are independent random variables, we have that the return follows a $\mathcal{G}_I^0$ distribution (Gambini et al., 2015; Naranjo-Torres et al., 2017), denoted $Z \sim \mathcal{G}_I^0(\alpha, \gamma, L)$, whose density is

$$f_Z(z; \alpha, \gamma, L) = \frac{L^L \Gamma(L - \alpha)}{\gamma^\alpha \Gamma(L)\Gamma(-\alpha)} \frac{z^L}{(\gamma + Lz)^{L-\alpha}}. \tag{4.11}$$

The k-order moment of Z is given by

$$\mathrm{E}Z^k = \left(\frac{\gamma}{L}\right)^k \frac{\Gamma(-\alpha - k)}{\Gamma(-\alpha)} \frac{\Gamma(L + k)}{\Gamma(L)}, \tag{4.12}$$

provided that $k < -\alpha$ and infinite otherwise.

4.2 Inference by Analogy

Inference by analogy (Manski, 1988) is inspired by the Law of Large Numbers, that states that (under relatively minor conditions) holds that

$$\lim_{n\to\infty} \frac{1}{n} \sum_{i=1}^{n} \psi(X_i) = \mathrm{E}\psi(X), \tag{4.13}$$

provided that $X, X_1, X_2, \ldots$ are independent identically distributed random variables.

With this in mind, and assuming one has a large sample, it seems reasonable to equate sample quantities (the left hand side) and parametric expressions (the right hand side).

When we have p parameters, i.e. $\boldsymbol{\theta} \in \Theta \subset \mathbb{R}^p$, we need p linearly independent equations to form an estimator of $\boldsymbol{\theta}$.

4.2.1 The Uniform Distribution

Using (4.4) we can set, for instance, $k = 1$ and obtain $\mathrm{E}U = \theta/2$. With this, our first-order moment estimator for θ is $\widehat{\theta}_1 = 2n^{-1}\sum_{i=1}^n U_i$. But we can also set $k = 2$ and obtain a second-order moment estimator, using $\mathrm{E}U^2 = \frac{1}{3}\theta^2$ it is immediate that $\widehat{\theta}_2 = \sqrt{3n^{-1}\sum_{i=1}^n U_i^2}$. In fact, the k-order moment estimator of θ based on a sample of size n is

$$\widehat{\theta}_k = \sqrt[k]{\frac{k+1}{n}\sum_{i=1}^{n} U_i^k}. \tag{4.14}$$

This multiplicity of possible analogy estimators gives great flexibility to the method, but it is also one of its weaknesses: the lack of unicity. Another drawback is that little is known about estimators obtained by this procedure, apart that they are consistent, i.e. that (4.13) grants that they converge in probability to the true parameter value.

A more serious problem is that an estimator obtained by this procedure may lead to a model for which observations are not feasible (remember, observations are always *correct*). See, for example, the sample $\boldsymbol{x} = (0.05, 0.15, 1)$. Its sample mean is $1.2/3 = 0.4$, therefore the estimate is $\widehat{\theta}_1 = 0.8$, but the third observation is unfeasible under the $\mathcal{U}_{(0,0.8)}$ distribution!

By the way, notice an important difference. We refer to an *estimator* when it is a random variable $\widehat{\theta}(X_1, X_2, \dots)$, and to an *estimate* when the data have been observed $\widehat{\theta}(x_1, x_2, \dots)$ and, thus, is a fixed quantity.

4.2.2 The Gaussian Distribution

Using $\mathrm{E}X = \mu$ we obtain $\widehat{\mu} = n^{-1}\sum_{i=1}^n X_i$. Since $\mathrm{E}X^2 = \mu^2 + \sigma^2$, we have that $\sigma^2 = \mathrm{E}X^2 - \mu^2$. We already have an estimator for μ, and also for μ^2, then one estimator for σ^2 is $\widehat{\sigma}^2 = n^{-1}\sum_{i=1}^n X_i^2 - \widehat{\mu}^2 = n^{-1}\sum_{i=1}^n (X_i - \widehat{\mu})^2$.

Other estimators could be formed with higher-order moments $k = 3, 4 \dots$, but we will always need two linearly independent equations to estimate $\boldsymbol{\theta} = (\mu, \sigma^2)$.

4.2.3 Mixture of Gaussian Distributions

We need $(n-1)n^2$ linearly independent equations to estimate $\boldsymbol{\theta} = (\boldsymbol{p}, \boldsymbol{\mu}, \boldsymbol{\sigma}^2)$. This approach is not very effective, because results in serious numerical problems.

4.2.4 The (SAR) Gamma Distribution

Remember that this distribution is characterized by the density presented in (4.9). Using $\mathrm{E}(Z) = \mu$, we propose $\widehat{\mu} = n^{-1}\sum_{i=1}^n Z_i$, and with $\mathrm{Var}(Z) = \mu^2/L$, we can estimate

$$\widehat{L} = \frac{\widehat{\mu}^2}{n^{-1}\sum_{i=1}^n (Z_i - \widehat{\mu})^2}. \tag{4.15}$$

This is the well-known estimator for the number of looks (equivalent number of looks) which consists of the reciprocal of the squared coefficient of variation. This estimator will be used in Section 6.3.3.

4.3 Inference by Maximum Likelihood

The principle of likelihood was formalized by R. A. Fisher. In most practical situations it produces unique estimators, and has good and well-known properties. It should be used whenever possible.

Consider again the sample independent of random variables $\boldsymbol{X} = (X_1, X_2, \dots, X_n)$ each with the same distribution characterized (without lack of generality) by the density $f(X_i; \boldsymbol{\theta})$. The likelihood function is

$$\mathcal{L}(\theta; \boldsymbol{X}) = \prod_{i=1}^{n} f(\theta; X_i). \tag{4.16}$$

Notice that $\mathcal{L}$ is not a joint density function, as it depends on θ, not on the variables.

The principle of maximum likelihood proposes as estimator for $\boldsymbol{\theta}$ the parameter that maximizes (4.16):

$$\widehat{\theta}_{\mathrm{ML}} = \arg\max_{\theta \in \boldsymbol{\Theta}} \{\mathcal{L}(\theta; \boldsymbol{X})\}, \tag{4.17}$$

that is, the point in $\boldsymbol{\Theta}$ that makes the observations most plausible. It sometimes coincides with some analogy estimators.

4.3.1 The Uniform Distribution

Assume $\boldsymbol{U} = (U_1, U_2, \dots, U_n)$ is a sample from $\mathcal{U}_{(0,\theta)}$, with $\theta > 0$ unknown. The likelihood is

$$\mathcal{L}(\theta; \boldsymbol{U}) = \prod_{i=1}^{n} \frac{1}{\theta} \mathbb{1}_{(U_i,\infty)}(\theta). \tag{4.18}$$

Denote the sorted sample in nondecreasing order $U_{1:n}, U_{2:n}, \dots, U_{n:n}$. After some manipulation, one arrives at the following expression:

$$\mathcal{L}(\theta; \boldsymbol{U}) = \frac{1}{\theta^n} \mathbb{1}_{(U_{n:n},\infty)}(\theta), \tag{4.19}$$

whose maximum is at $U_{n:n}$, so $\widehat{\theta}_{\mathrm{ML}} = \max\{U_1, U_2, \dots, U_n\}$.

This is quite different from any analogy estimator, and the estimator by maximum likelihood is always admissible.

4.3.2 The Gaussian Distribution

In this case, $\widehat{\theta}_{\mathrm{ML}}$ coincides with the analogy estimators we already derived using the analogy approach.

4.3.3 Mixture of Gaussian Distributions

The ML estimator poses a difficult optimization problem. It consists of finding

$$\widehat{(\boldsymbol{p}, \mu, \sigma^2)}_{\mathrm{ML}} = \arg \max_{(\boldsymbol{p},\mu,\sigma^2)\in\boldsymbol{\Theta}} \sum_{j=1}^{n} \frac{1}{\sqrt{2\pi}\sigma_j} \exp\left\{ -\frac{1}{2\sigma_j^2}(x-\mu_j)^2 \right\}. \tag{4.20}$$

This is so difficult and numerically unstable, that it is often solved by using the EM (Expectation-Maximization) algorithm. It is highly recommended to use the `mclust` package (Farley et al., 2012) available in `R` (R Core Team, 2020).

Since we assume no prior knowledge about the number of components n, it is important to have a balance between the number of parameters and the likelihood, otherwise we may end up with a model with the largest possible n.

There are many measures of the quality of models, among them BIC – Bayesian Information Criterion, and AIC – Akaike Information Criterion. The latter is defined as

$$\mathrm{BIC} = \#\boldsymbol{\Theta} \ln n - 2 \ln \mathcal{L}(\widehat{\theta}; \boldsymbol{X}),$$

and the preferred model is the one that minimizes this quantity.

In the following, we will see the analysis of the `rms_hist75` data set, which consists of 10628 radiometry observations. Figure 4.2 shows the estimated densities of a Gaussian distribution (solid line), along with Gaussians models estimates with $n = 2, 3, 4, 5$ components (dashed, dotted, dot-dash, long dashes, and two dashes, respectively). The larger the number of components, the closer the fit to the data.

4.3.4 The (SAR) Gamma Distribution

The likelihood function is

$$\begin{aligned} \mathcal{L}(L, \mu; \boldsymbol{Z}) &= \prod_{i=1}^{n} f_\Gamma(L, \mu; Z_i) = \prod_{i=1}^{n} \frac{L^L}{\mu^L \Gamma(L)} Z_i^{L-1} \exp\{-LZ_i/\mu\} \\ &= \left(\frac{L^L}{\mu^L \Gamma(L)}\right)^n \prod_{i=1}^{n} [Z_i^{L-1} \exp\{-LZ_i/\mu\}]. \end{aligned} \tag{4.21}$$

This is a very tough function to maximize.

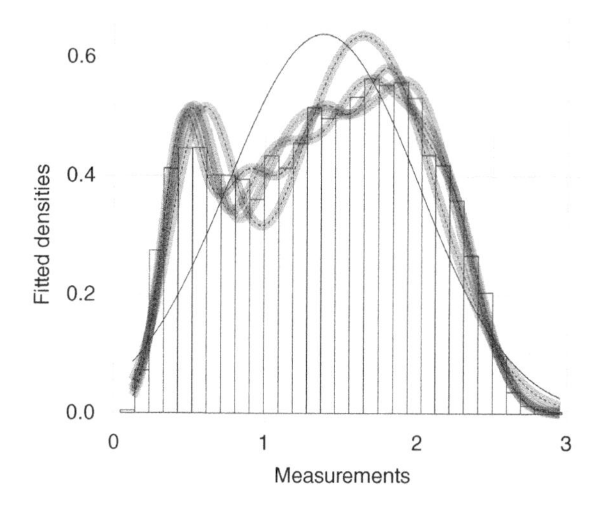

Figure 4.2 Histogram, individual components, and resulting mixture model.

A usual trick is taking the logarithm, as the likelihood function is positive. One then obtains the "complete log-likelihood function":

$$\ell^*(L,\mu;\boldsymbol{Z}) = n\left[L(\log L - \log\mu) - \log\Gamma(L)\right] + (L-1)\sum_{i=1}^{n}\log Z_i - \frac{L}{\mu}\sum_{i=1}^{n} Z_i. \tag{4.22}$$

But there are terms that do not depend on either L or μ and, therefore, are irrelevant for our maximization problem. We are finally interested in the "reduced log-likelihood function":

$$\ell(L,\mu;\boldsymbol{Z}) = n\left[L(\log L - \log\mu) - \log\Gamma(L)\right] + L\sum_{i=1}^{n}\log Z_i - \frac{L}{\mu}\sum_{i=1}^{n} Z_i. \tag{4.23}$$

The maximum likelihood estimator of (L,μ) is then any point in $\mathbb{R}_+^2$ satisfying

$$\widehat{(L,\mu)} = \arg\max_{(L,\mu)\in\mathbb{R}_+^2} \ell(L,\mu;\boldsymbol{Z}). \tag{4.24}$$

This problem can be solved in two ways: either by deriving, equating to zero and solving, or by direct maximization. Each case must be studied from a computational viewpoint to choose the most suitable option.

4.3.5 The $\mathcal{G}^0$ Distribution

Recall that the $\mathcal{G}^0$ distribution was proposed as a means to model data from textured and extremely textured areas. Assume that some textureless areas have been identified and that, with these data, it was possible to obtain a dependable estimate of L, say $\widehat{L}$. As this equivalent number of looks should be valid for the whole image, we will consider it know when fitting data from textured areas.

Assume, then, that we have a sample $\boldsymbol{Z} = (Z_1,\dots,Z_n)$ of iid (independent identically distributed) random variables that follow the $\mathcal{G}^0(\alpha,\gamma,\widehat{L})$ distribution. The unknown parameter $\theta = (\alpha,\gamma)$ lies in $\Theta = \mathbb{R}_- \times \mathbb{R}_+$.

The maximum likelihood estimator of (α,γ), is any point that maximizes the reduced log-likelihood:

$$\ell(\alpha,\gamma,\widehat{L},\boldsymbol{Z}) = \log\frac{\Gamma(\widehat{L}-\alpha)}{\gamma^{\alpha}\Gamma(-\alpha)} + \widehat{L}\sum_{i=1}^{n}\log\frac{Z_i}{\gamma+\widehat{L}Z_i} + \alpha\sum_{i=1}^{n}\log(\gamma+\widehat{L}Z_i), \tag{4.25}$$

provided it lies in Θ. Maximizing (4.25) might be a difficult task, in particular in textureless areas where $\alpha\to-\infty$ and the $\mathcal{G}^0$ distribution becomes very close to a Gamma law. Small samples also pose difficult numerical problems, as ℓ becomes flat (Frery et al., 2004).

Most algorithms for maximizing (4.25) require a starting point. A good solution consists in using estimates obtained with the method of moments, i.e. by forming a system of two equations with suitable different values of k in (3.13).

4.4 Analogy vs. Maximum Likelihood

These methods should not be seen as competitors; they are complementary.

Analogy is usually more straightforward than Maximum Likelihood. In particular, it does not require the expression of the density of the model (think, for instance, in the problem of estimating the parameter of the sum of k random variables with $\mathcal{U}_{(0,\theta)}$ distribution). Analogy leads to finding the roots of a system of (usually nonlinear) equations, and this is usually cumbersome in high-dimensional parametric spaces.

Estimators derived by Maximum likelihood have many asymptotic properties, and they are regarded as the best ones for large samples when there is no contamination. They can be obtained by either finding the roots of a system of (again, usually nonlinear) equations, which shares the problems of Analogy, or by optimization of the reduced log-likelihood function. There are a number of excellent algorithms for the latter approach, numerical optimization (Henningsen and Toomet, 2011), EM and Simulated Annealing among them.

One frequently uses an analogy estimate as the starting point for optimization techniques that seek the maximum likelihood estimate.

4.5 Improvement by Bootstrap

Bootstrap is a resampling technique. We will use its simplest version.

Consider you have $\widehat{\theta}$, an estimator of θ based on the sample $\boldsymbol{X} = (X_1, \dots, X_n)$.

Its bias is $B(\widehat{\theta}) = \mathrm{E}\widehat{\theta} - \theta$.

A better estimator would be, in terms of bias,

$$\begin{aligned} \dot{\theta} &= \widehat{\theta} - B(\theta) && (4.26) \\ &= \widehat{\theta} - \mathrm{E}\widehat{\theta} + \theta \\ &= \widehat{\theta} + \theta - \mathrm{E}\widehat{\theta}, && (4.27) \end{aligned}$$

but we know neither θ nor $\mathrm{E}\widehat{\theta}$.

What do we do, as statisticians, when we do not know a quantity? We estimate it! So we propose the following estimator

$$\begin{aligned} \tilde{\theta}(\boldsymbol{X}) &= \hat{\hat{\theta}} = \widehat{\theta} + \widehat{\theta} - \widehat{\mathrm{E}\widehat{\theta}} \\ &= 2\widehat{\theta} - \frac{1}{R}\sum_{r=1}^{R}\widehat{\theta}(\boldsymbol{X}^{(r)}), && (4.28) \end{aligned}$$

where $\boldsymbol{X}^{(r)}$ is the result of resampling $\boldsymbol{X}$ with replacement.

In spite of seeming naive and ad hoc, it has a solid theoretical foundation and, more often than not, the bootstrap estimator is excellent, specially for relatively small samples (Cribari-Neto et al., 2002, Vasconcellos et al., 2005, Silva et al., 2008).

Notice that the number of possible permutations with repetitions of the sample $\boldsymbol{X} = (X_1, \dots, X_n)$ is n^n. Therefore, it is important to verify which is smaller: if n^n or R. If $n^n < R$, then there is no need to sample from the n^n permutations; it is possible to make a deterministic bootstrap, with all of them.

A final warning: it is important to know how the estimator $\widehat{\theta}$ is computed. If it involves dividing by the variance or the standard deviation, we cannot use those bootstrap samples with the same observation, otherwise we will have a division by zero.

4.6 Comparison of Estimators

Assume you have two estimators, say $\widehat{\theta}$ and $\tilde{\theta}$. Denote either of them by $\dot{\theta}$. They can be compared according to:

- their bias $B(\dot{\theta}, \theta) = \mathrm{E}\dot{\theta} - \theta$,
- their variance $\mathrm{Var}\dot{\theta}$,
- their mean quadratic error $\mathrm{MQE}(\dot{\theta}) = \mathrm{E}(\dot{\theta} - \theta)^2$.

Also to be considered:

- are they fast and easy to compute?,
- how fast do they converge to the true value?,
- do they converge, and how fast, to a Gaussian distribution?, and
- are they robust before different types of contamination?

See details in Bustos and Frery (1992). Variance and bias should be as small as possible. They must always be checked, either analytically (the ideal scenario), or by a well-designed Monte Carlo study.

4.7 An Example

Consider the Exponential distribution, a particular case of the (SAR) Gamma law characterized by the density given in (4.9). Making $L = 1$ one has

$$f_Z(z; \mu) = \frac{1}{\mu} \exp\{-z/\mu\}, \tag{4.29}$$

with $\mu > 0$. We denote this situation $Z \sim E(\mu)$. Recall Chapter 3, where we defined it in (3.3). Figure 3.2 illustrates this density.

Consider the sample $\boldsymbol{Z} = (Z_1, \dots, Z_n)$ of independent random variables such that $Z_i \sim \mathrm{Exp}(\mu)$. The reader is invited to verify that both the maximum likelihood and first moment estimators are the sample mean:

$$\widehat{\mu} = \frac{1}{n}\sum_{i=1}^{n} Z_i. \tag{4.30}$$

We may obtain another analogy estimator noting that the median of Z_i is $Q_{1/2}(Z_i) = \mu \ln 2$, therefore we may use

$$\breve{\mu} = \frac{1}{\ln 2} q_{1/2}(z_1, \dots, z_n), \tag{4.31}$$

as an estimator, where $q_{1/2}(\boldsymbol{z})$ denotes the sample median. Since we know that maximum likelihood estimators are optimal with large samples from the hypothesized model, we may consider an improvement of $\breve{\mu}$, its bootstrapped version $\tilde{\mu}$.

With this, we have three estimators, namely $\widehat{\mu}$, $\breve{\mu}$, and $\tilde{\mu}$. We will now compare them with a Monte Carlo experiment.

A well-designed Monte Carlo experiment consists of stipulating, at least, the following elements:

1. a model, in our case $Z_1, \dots, Z_n$ independent random variables such that $Z_i \sim E(\mu)$; we may extend this model by considering, for instance, several types of contamination;
2. the relevant factors which, for us, are the sample size n and the mean μ; we could have added the number of replications R in the bootstrap. We will fix it as $R = 300$.

The relevant factors (item 2) impose the size of the data structure where we will collect the observations. Assume we will analyze $n \in \{3, 5, 10, 20, \dots, 100, 1000, 10000\}$ and, for the sake of simplicity, $\mu = 1$, then our factor space size is $14 \times 1 = 14$. For each point in the factor space we will obtain the bias, and the mean quadratic error of each estimator, i.e. for each of the 14 sample sizes we will collect six observations. We will store these observations as a 14×7 matrix: one line for each sample size n, and the columns being the sample size n, followed by the sample bias and mean square error. See

Notice that $3^3 = 27$ and $4^4 = 256$ are smaller than $R = 300$, and that $n^n > 300$ for every $n \geq 5$. So, if we want to make an efficient Monte Carlo study, we should opt for using all the n^n permutations with repetitions in these two cases.

Bustos and Frery (1992) suggest using a variable number of replications for each sample size in problems like this one. Their proposal aims at controlling the variability of the estimates of, e.g. the bias and the mean quadratic error, making them comparable within the study. We will follow this advice, and use $r = \lceil N/n \rceil$, Since our largest sample is $n_{\text{max}} = 10^4$, and we would like 200 replications in this case, we set $N = 2 \cdot 10^6$. With this, the maximum number of replications is $r_{\text{max}} = 666667$, corresponding to $n_{\text{min}} = 3$.

The following code implements this Monte Carlo experiment.

Line 3 is a fundamental step in every simulation study: setting the seed. This grants reproducibility, as every user will obtain the same results. Always check the documentation for good seeds. Lines 30 and 61 store the initial and final times, used in line 62 to exhibit the total running time. Line 5 starts the definition of the median estimator improved by bootstrap. Notice that the number of bootstrap replications is a parameter. Lines 11 to 15 define the deterministic bootstrap. The definition ends at line 25.

Line 27 defines the factor space of our study, namely the set of all sample sizes. Line 28 builds space for the output. Line 32 defines the variable which stores the position in the factor space; it will be used to index the matrix defined in the previous line.

Line 33 starts the main loop; the variable `n` will access each value of the factor space, which is stored in `N`.

Line 41 begins the loop for the current sample size. Line 36 stores in `r` the current sample size. Lines 37, 38, and 39 create space for storing the `r` observations of each estimator. The loop ends in line 47.

When this inner loop ends, the variables `v.mu1`, `v.mu2` and `v.mu3` store the `r` observations of each estimator. We then calculate the bias and mean square error of each (lines 49 and 50 for $\widehat{\mu}$, and the following four for $\breve{\mu}$ and $\tilde{\mu}$). Notice that three estimators are computed with the same sample. This is the output of the current state of `r`, the sample size. These values are stored in line 58, and the inner loop restarts with the next sample size, or ends if it reached the end.

This code runs in approximately 2 h 45 min in a MacBook Pro 3.5 GHz Intel Core i7 with 16 GB of memory, running `R` version 4.1.1 (Kick Things) on MacOS Big Sur version 11.5.2.

> Important notice: Never, ever, use a machine state to initialize the pseudorandom number generator.

Figure 4.3 shows the data we produced and stored in `BiasMSE`: the estimated dependence of the bias on the sample size (Figure 4.3a), and the estimated dependence of the mean square error on the sample size (Figure 4.3b).

Listing 4.1: Monte Carlo code

```
1  require(gtools)

3  set.seed(1234567890)

5  est.median.bootstrap <- function(z, R) {
6      t <- median(z) / log(2)

8      sample_size <- length(z)
9      pwr <- sample_size^sample_size

11     if(pwr < R) {
12         m.Bootstrap <- permutations(sample_size, sample_size, z,
13         set=TRUE, repeats.allowed=TRUE)
14         return(2*t - mean(unlist(lapply(m.Bootstrap, median))))
15     }
16     else {
17         v.Bootstrap <- rep(0, R)
18         for(b in 1:R) {
19             x <- sample(z, replace = TRUE)
20             v.Bootstrap[b] <- median(x) / log(2)
21         }
22     }

24     return(2*t - mean(v.Bootstrap))
25 }

27 N <- c(3, 5, seq(10,100,by=10), 1000, 10000)
28 BiasMSE <- matrix(nrow=14, ncol=7)

30 start_time <- Sys.time()

32 i <- 0
33 for(n in N){
34     i <- i+1

36     r <- ceiling(2*10^6/n)
37     v.mu1 <- array(rep(0, r))
38     v.mu2 <- array(rep(0, r))
39     v.mu3 <- array(rep(0, r))

41     for(j in 1:r){
42         z <- rexp(n) # sample of size n from Exp(1)

44         v.mu1[j] <- mean(z)
45         v.mu2[j] <- median(z) / log(2)
46         v.mu3[j] <- est.median.bootstrap(z, 300)
47     }

49     bias1 <- mean(v.mu1) - 1
50     eqm1 <- mean((v.mu1 - 1)^2)

52     bias2 <- mean(v.mu2) - 1
53     eqm2 <- mean((v.mu2 - 1)^2)

55     bias3 <- mean(v.mu3) - 1
56     eqm3 <- mean((v.mu3 - 1)^2)

58     BiasMSE[i,]  <- c(N[i], bias1, eqm1, bias2, eqm2, bias3, eqm3)
59 }

61 end_time <- Sys.time()
62 end_time - start_time
```

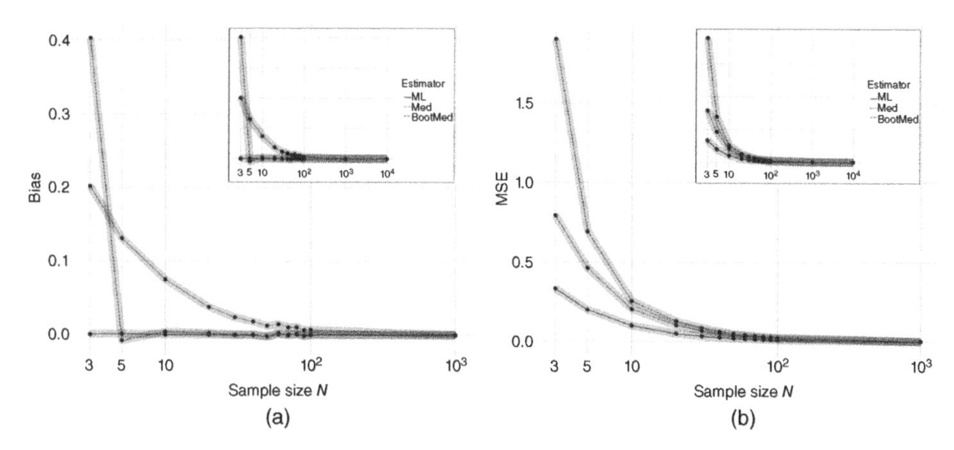

Figure 4.3 (a) Bias and (b) mean square error of $\widehat{\mu}$ (ML), $\breve{\mu}$ (Med), and $\widetilde{\mu}$ (BootMed).

From Figure 4.3 we may conclude, among other things, that under the current model:

- There seems to be no need to consider samples of size $n = 10000$ or larger, since the bias of the three estimators is very close with $n = 1000$, and the same holds for the mean square error even for smaller samples.
- The bias of $\widehat{\mu}$ is almost constant for every sample size, while the bias of $\breve{\mu}$ starts large, and decreases with the sample size.
- The effect of bootstrap over the median estimator is evident with samples of at least size $n = 5$; the case $n = 3$ is hopeless.
- Regarding the bias only, the bootstrap-corrected median and maximum likelihood estimators behave alike with samples of at least five observations.
- The mean square error shows a very smooth decrease with respect to the sample size in the three estimators, but with different scales. The smallest MSE is, consistently, due to $\widehat{\mu}$, followed by $\breve{\mu}$, and by $\widehat{\mu}$.
- If the only criterion is the mean square error, the safest choice is the maximum likelihood estimator.

So, why one should consider estimators other than maximum likelihood? Because this family of estimators is know to have the best performance under the hypothesized model, but they may lead to catastrophic results when the observations are contaminated.

The reader is referred to the textbooks by Huber (1981), Huber and Ronchetti (2009), and Maronna et al. (2006). Bustos et al. (2002) and Allende et al. (2006) who discuss the use of robust estimators in the analysis of SAR data. Section 6.3 is devoted to reviewing topics of robustness in SAR data analysis.

4.8 The Same Example, Revisited

So, what is *robustness* about?

Every statistical analysis relies on explicit or implicit assumptions. The most widespread of these is the Gaussian distribution, followed by the independence. Estimators are tailored to the assumptions, and if the underlying system departs from them, the consequences may be catastrophic.

Imagine, for instance, that the observations are being registered by hand on a spreadsheet. Even if the process is careful, it is prone to errors, e.g. missing or misplacing the decimal point.

If all possible errors could be described with precision, a careful statistician could, in principle, include them in the model. This would, in principle, solve the problem although it may lead to very complicated estimators. But, in practice, the analyst can hardly foresee all types of errors.

Then, robustness comes in. The idea of devising robust procedures is, essentially, pessimistic, but in a good sense. A robust estimator should be prepared to handle several types of departures from the classical or basic hypotheses, but not at the cost of being too complicated or too ineffective when, alas, the data obey them. We will go back to this topic with more details in Section 6.3.

In this example we will see the effect of contamination on the three estimators we assessed in Section 4.7. We will consider that each observation may or may not be contaminated by a factor of 100 (missing the decimal point would have this kind of effect) with a very small probability $\varepsilon = 10^{-3}$ (our operator misses, in mean, one every thousand decimal points).

Our model becomes

$$Z_i \sim \begin{cases} \text{Exp}(\mu) & \text{with probability } 1-\varepsilon, \\ \text{Exp}(100\mu) & \text{with probability } \varepsilon. \end{cases} \tag{4.32}$$

This is a mixture model, but instead of treating it as in Section 4.2.3, we will see how the estimators presented in Section 4.7 behave.

The only change required is the line that produces the samples, cf. line 42, by the following:

```
z <- rexp(n) * 100^rbernoulli(n,.001)
```

The results are presented in Figure 4.4.

The most striking difference between the results presented in Figures 4.3 and 4.4 is the behavior of $\widehat{\mu}$, the maximum likelihood estimator. Whereas with "pure" samples it was consistently the best estimator in terms of bias and mean square error, it becomes the worst one with a relatively small departure from the hypothesis of "pure" data. The maximum likelihood estimator is consistently biased, and it has the largest mean square error. In terms of bias, the bootstrap-improved median is the best option, and it pairs the mean square error of the median with samples of at least ten observations.

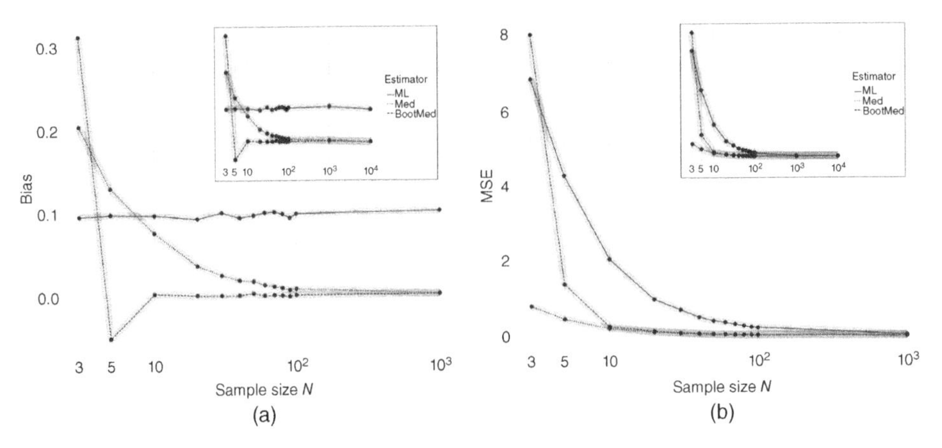

Figure 4.4 (a) Bias and (b) mean square error of $\widehat{\mu}$ (ML), $\check{\mu}$ (Med), and $\tilde{\mu}$ (BootMed) with contaminated data.

This example concludes with a general recommendation: whenever possible, use two estimators, namely maximum likelihood and a robust one. If they provide significantly different results, consider making a deep analysis of the data, revise your hypotheses, and, ultimately, rely on the robust one for your subsequent interpretation.

4.9 Another Example

We will conclude this chapter with the analysis of data from an urban area. Figure 4.5 shows, to the left, a 200×300 pixel sample from a densely urbanized area, as seen in the HV polarization, single look. We will analyze the smaller sample to the right, which consists of 111×51 pixels.

The summary of this sample is as follows, showing (as expected) a great deal of variability.

```
> summary(vUrbanHV)
UHV
Min.    :      0
1st Qu.:   7156
Median :  22345
Mean    :  69428
3rd Qu.:  69577
Max.    :3266496
```

The usual histogram (to the left) is barely informative, as a few large observations make all the rest concentrate to the left; cf. Figure 4.6. The bin width (w) was calculated with the Freedman-Diaconis rule-of-the-thumb: $w = 2n^{-1/3}\text{IQR}(\mathbf{z})$, where IQR denotes the inter-quartile range, and n is the number of observations (Freedman and Diaconis, 1981). When we restrict the histogram to those values below 10^5 (right), we notice an exponential-like shape.

We will use the result given in (4.12) to build an estimator based on the first and second moments. With this, one has that if $Z \sim \mathcal{G}^0(\alpha, \gamma, L)$, then

$$\mathrm{E}(Z) = \frac{\gamma}{-\alpha - 1}, \text{ and} \tag{4.33}$$

$$\mathrm{E}(Z^2) = \frac{\gamma^2}{L} \frac{L+1}{(-\alpha - 1)(-\alpha - 2)}. \tag{4.34}$$

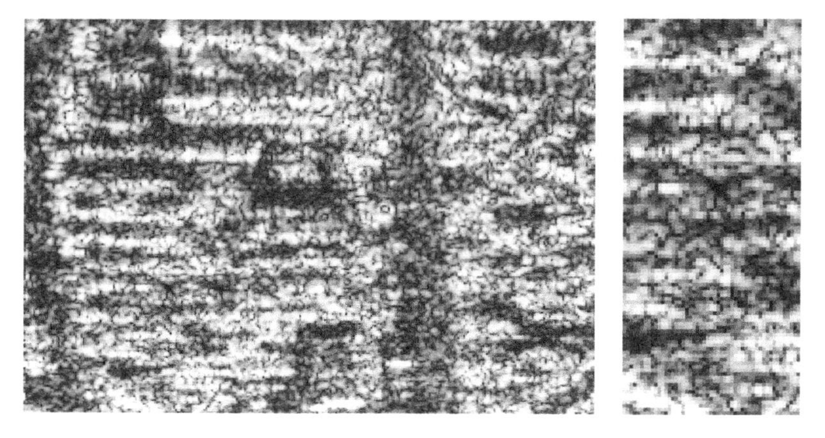

Figure 4.5 Urban area after equalization, and smaller sample under analysis.

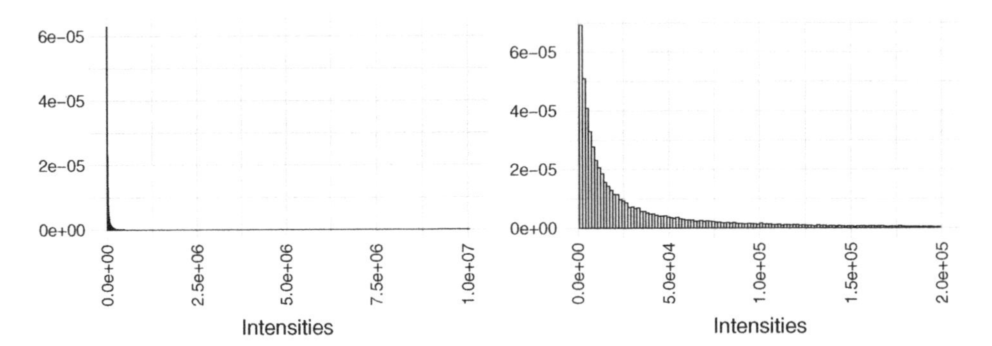

Figure 4.6 Histograms of data from the urban area: full and restricted.

Now using the fist and second sample moments, m_1 and m_2, respectively, and equating with (4.33) and 4.34 indexed by estimators of α and γ, $\breve{\alpha}$ and $\breve{\gamma}$, resp., we obtain

$$m_1 = \frac{\breve{\gamma}}{-\breve{\alpha} - 1}, \text{ and} \tag{4.35}$$

$$m_2 = \frac{\breve{\gamma}^2}{L} \frac{L+1}{(-\breve{\alpha} - 1)(-\breve{\alpha} - 2)}. \tag{4.36}$$

With this, we may obtain the estimators as

$$\breve{\alpha} = -2 - \frac{L+1}{L\, m_2/m_1^2}, \text{ and} \tag{4.37}$$

$$\breve{\gamma} = m_1 \left(2 + \frac{L+1}{L\, m_2/m_1^2} \right). \tag{4.38}$$

The following code implements this estimator. It requires the sample and the number of looks as input, and provides the estimators with their names as a list.

```
1         GI0.Estimator.m1m2 <- function(z, L) {
2                 m1 <- mean(z)
3                 m2 <- mean(z^2)
4                 m212 <- m2/m1^2

6                 a <- -2 - (L+1) / (L * m212)
7                 g <- m1 * (2 + (L+1) / (L * m212))

9                 return(list("alpha"=a, "gamma"=g))
10        }
```

Using the urban data we are analyzing, we obtain $\breve{\alpha} = -2.208010$ and $\breve{\gamma} = 100105.90$.

We already have in (4.25) the expression of the maximum likelihood estimator for (α, γ). The following code shows its implementation and usage with the function `maxNR` from the `maxLik` package (Henningsen and Toomet, 2011). Notice that use the moments estimates as starting points for the algorithm, and we assume $L = 1$ known.

```
1         LogLikelihoodLknown <- function(params) {

3                 p_alpha <- -abs(params[1])
4                 p_gamma <- abs(params[2])
5                 p_L <- abs(params[3])
```

```
7             n <- length(z)

9             return(
10            n*(lgamma(p_L-p_alpha) - p_alpha*log(p_gamma)
11               - lgamma(-p_alpha)) +
12            (p_alpha-p_L)*sum(log(p_gamma + z*p_L))
13            )
14        }

16        z <- vUrbanHV$UHV
17        estim.UrbanML <- maxNR(LogLikelihoodLknown,
18        start=c(estim.Urban$alpha, estim.Urban$gamma, 1),
19        activePar=c(TRUE,TRUE,FALSE))$estimate[1:2]
```

With this approach, we obtain $\widehat{\alpha} = -2.662928$ and $\widehat{\gamma} = 62808.46$.

Table 4.1 shows these estimates. Notice that the largest differences are observed in the estimates of γ.

Figure 4.7 shows the restricted histogram of the urban data, with the fitted densities of the Exponential and $\mathcal{G}^0$ densities. The latter is shown with the maximum likelihood and

Table 4.1 Estimates for the Urban area with $L = 1$.

	α	γ
Moments	−2.208010	100105.90
Maximum likelihood	−2.662928	62808.46
Bootstraped ML	−2.460509	36841.35

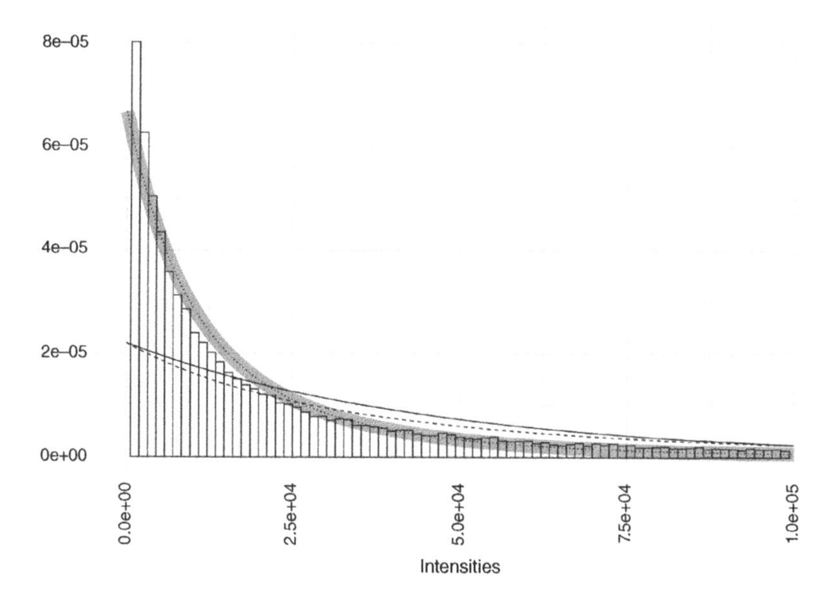

Figure 4.7 Restricted histogram of the urban area data, with fitted exponential (solid) and $\mathcal{G}^0$ densities; the latter is shown with the parameters estimated by maximum likelihood (dashed) and its improved version with bootstrap (dots).

bootstrap-improved maximum likelihood estimates. In spite of the numerical difference between the moments and maximum likelihood estimates, cf. Table 4.1, the corresponding estimated densities are not distinguishable on the plot.

On the one hand, both the Exponential and $\mathcal{G}^0$ densities fail at fitting the data on the restricted domain shown here when they are indexed by estimates of the maximum likelihood estimators. On the other hand, when the maximum likelihood estimator of the parameters of the $\mathcal{G}^0$ distribution is improved by bootstrap, we obtain a much better fit.

Exercises

1 Consider $\boldsymbol{U} = (U_1, \dots, U_n)$, a sample of iid $\mathcal{U}_{(0,\theta)}$ random variables with $\theta > 0$ unknown. Obtain at least three estimators for θ by the method of moments. Compare them, in terms of bias and mean quadratic error, with respect to the maximum likelihood estimator. Use a variety of sample sizes and true values of θ.

2 Improve all estimators obtained in Exercise 1 by the bootstrap method. Experiment with different bootstrap replications. Consider also execution times.

3 Consider the model given in (4.8) such that $p = 2$, $\mu_1 = -1$, $\mu_2 = 1$, $\sigma_1^2 = \sigma_2 = 1$, and $p_1 = 1/2$. Write a routine for sampling from this model. Now work with three situations of unknown means and variances: (i) one component, (ii) two components, and (iii) three components. Assess the BIC of each situation by means of a Monte Carlo experiment, varying the sample size; you may use 1000 replications for each situation.

4 Compare the maximum likelihood and a moment estimator for the model given in (4.9) by means of a Monte Carlo experiment, using the bias and the mean quadratic error as measures of quality. Consider a variety of true parameters and sample sizes.

5 Extend Exercise 4 by including bootstrap-improved estimators. Make a global comparison.

6 Find at least two estimators for the model given by (3.9). Compare them for a variety of parameters.

7 Extend Exercise 6 by including bootstrap-improved estimators. Make a global comparison.

5

Applications

Despeckling filters based on statistical methods for Synthetic Aperture Radar (SAR) imagery have been chosen to put into practice the statistical models studied due to their relevance in the analysis and interpretation of SAR images. Among those filters, we discuss the ones based on the Mean, the Median, and the Lee filters that stem from this statistical modeling. Then, we comment the Maximum a posteriori (MAP) filter, and the nonlocal approach (the original and the state-of-the art statistical nonlocal filters using stochastic distances and hypothesis test). These despeckling filters are commonly used in SAR despeckling, providing excellent results. For each filter, a brief introduction is first addressed, and then, some applications for both, simulated and actual SAR data are given. Image classification has also been addressed through the standard (classical) elemental machine learning methods. Many examples are shown and also, all the codes are available at the web site www.wiley.com/go/frery/sarimageanalysis.

5.1 Statistical Filters: Mean, Median, Lee

Given the coherent imaging mechanism used in the SAR system, speckle is always accompanied by SAR images, which causes a dilemma for further image processing, such as segmentation, classification, etc. Thus, speckle reduction is one necessary step for SAR image processing. Furthermore, given the randomness properties of the speckle in SAR images, the statistics-based filter is widely used for SAR image despeckling.

This section introduces three classical filtering methods, including Mean filter, Median filter, and Lee filter. Note that all these filters are designed according to local statistics. Due to the use of statistics, these filters work well under some particular assumptions. Thus, better performance will be obtained by these filters when the corresponding assumptions are well satisfied. In Sections 5.1.1, 5.1.2, and 5.1.3, we will describe these three filters in detail. Moreover, further analysis will also be made using the simulated and the actual noise images (including authentic SAR images).

5.1.1 Mean Filter

The Mean filter is widely used for image processing, especially with optical images, as it is the simplest linear filter. Note that, the Gaussian additive noise model is usually used in optical images. Since the average value computed from a local smaller window is used as the output

SAR Image Analysis — A Computational Statistics Approach: With R Code, Data, and Applications, First Edition.
Alejandro C. Frery, Jie Wu, and Luis Gomez.

Companion website: www.wiley.com/go/frery/sarimageanalysis

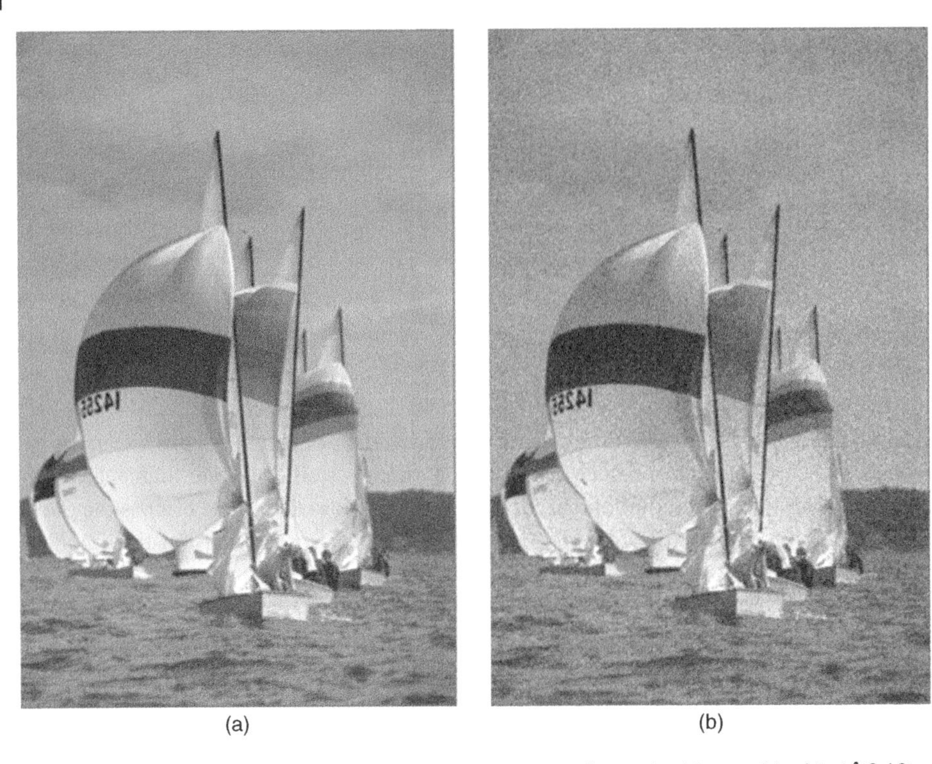

Figure 5.1 An examples of optical image (a) and corresponding noised image (b) with $\mathcal{N}(0,10)$.

of the filter, the random noise's mean is postulated to be zero for the Mean filter. Thus, if the assumption of random noise's mean equals zero is not satisfied, a biased estimation will be obtained.

Here, using an optical image (named as boat, nested in the R package, shown in Figure 5.1(a)) as an example, Gaussian additive noise $\mathcal{N}(0,10)$ is added as shown in Figure 5.1(b). For the Mean filtering, local windows of three different sizes are used, namely 3×3, 5×5, and 7×7. The visual results are given in Figure 5.2.

Figure 5.2 Denoised images with the mean filter of (a) 3×3 window; (b) 5×5 window; (c) 7×7 window.

The corresponding R codes are available in file `Code/R/Chapter6/611.R` from www.wiley.com/go/frery/sarimageanalysis, and shown as follows:

```
rm(list = ls())
source("../imagematrix/imagematrix.R")
library(imager)
library(plot3D)
## original image saving
boats.gray <- grayscale(imrotate(boats, -90))
boats.gray <- as.cimg(boats.gray[1:383, 1:256])
imagematrixPNG(imagematrix(boats.gray, "grey"), "boats.png")
boats.dim <- dim(boats.gray)
## Gaussian noise adding
boats.noise <-
boats.gray * 255 + imager::imnoise(boats.dim[1], boats.dim[2], sd = 10)
mse <- mean((boats.gray * 255 - boats.noise) ^ 2)
print(mse)
## scaling the range of noise image to [0,1]
boats.noise.normal <- boats.noise / 255
## noise image showing and saving
plot(imagematrix(boats.noise.normal, "grey"))
imagematrixPNG(imagematrix(boats.noise.normal, "grey"),
"boats_noise_Gauss.png")
noise.type <- "Gauss"
## denosing process
for (windowsize in c(3, 5, 7)) {
        xy <- pixel.grid(boats.noise)
        boats.noise.patches <-
        extract_patches(boats.noise,
        xy$x,
        xy$y,
        windowsize,
        windowsize,
        boundary_conditions = 3L)
        ## mean filtering
        denoise.type <- "mean"
        boats.denoise <- list()
        for (i in 1:length(boats.noise.patches)) {
                boats.denoise <-
                c(boats.denoise, mean(unlist(boats.noise.patches[[i]])))
        }
        boats.denoise.img <-
        matrix(unlist(boats.denoise), boats.dim[1], boats.dim[2])
        ## denoised image showing and saving
        boats.denoise.normal <- boats.denoise.img / 255
        plot(imagematrix(boats.denoise.normal, "grey"))
        filename <-
        paste("boats", noise.type, denoise.type, windowsize, sep = "_")
        filename <- paste(filename, "png", sep = ".")
        imagematrixPNG(imagematrix(boats.denoise.normal, "grey"), filename)
        mse <- mean((as.matrix(boats.gray * 255) - boats.denoise.img) ^ 2)
        print(mse)
}
```

Visually, a larger window is good at noise reduction than a smaller window, while the details are more prone to be smoothed for a larger window. Thus, window size should be adjusted for better results. To measure the difference between the original and the denoised images, mean square error (MSE) is computed for each window. The results are shown in Table 5.1. Note that the mean square error between the original and the noised images is 99.7707. This means, using a 3×3 window, a minor deviation is obtained while a more significant departure is produced by using a larger window.

5.1.2 Median Filter

As the simplest nonlinear filter, the Median filter uses the median value of the local window as the local estimation (Ng and Ma, 2006). This means that the local median value is supposed to be closer to the actual value than other values. Thus, the random noise values have equal

Table 5.1 MSE of different windows' size.

Window size	MSE
3×3	70.5011
5×5	151.3596
7×7	209.9131

probability to lie on either sides of the actual value, such as Gaussian additive noise $\mathcal{N}(0, \sigma)$ and Impulse noise.

Here, using an optical image (shown in Figure 5.1(a)) as an example, Gaussian additive noise $\mathcal{N}(0,10)$ is also added. The results of the Median filter are given in Figure 5.3. As the median value is estimated from a local window, the window's size is also important for the performance of the Median filter. Similar to the Mean filter, three windows are used in Figure 5.3, including 3×3, 5×5, and 7×7. Since the corresponding R code is very similar to that of the Mean filter, the codes are not given here but available in file `Code/R/Chapter6/612.R` from www.wiley.com/go/frery/sarimageanalysis.

For the visual results shown in Figure 5.3, we can see that the noise is suppressed more strongly by using a larger window while the details are smeared at the same time. This is the same behavior as that of the Mean filter. Thus, we should use a suitable window size for better performance. Using MSE as a measure, Table 5.2 gives the corresponding values of the Median filter of three different windows.

With a careful comparison between Tables 5.1 and 5.2, we can see that the Median filter obtains a lower MSE with the same window.

Given the salt and pepper property of impulse noise that is different from Gaussian additive noise, the optical image is corrupted with impulse noise and shown in Figure 5.4 (the ratio is 0.2). Using the Mean filter and Median filter, the corresponding visual results of different windows are shown in Figure 5.5. Also, the result for MSE is computed and listed in

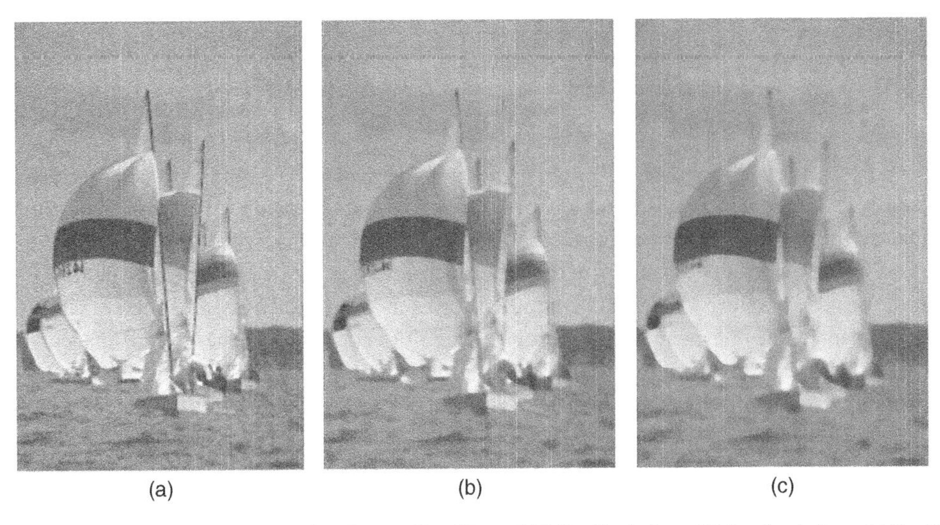

(a) (b) (c)

Figure 5.3 Denoised images with the median filter of (a) 3×3 window; (b) 5×5 window; (c) 7×7 window.

Table 5.2 MSE of different windows' size for Gaussian additive noise.

Window size	MSE
3×3	58.1288
5×5	145.8041
7×7	200.9034

Figure 5.4 Images corrupted with impulse noise (ratio is 0.2).

Table 5.3. The R codes are available in file `Code/R/Chapter6/613.R` from www.wiley .com/go/frery/sarimageanalysis. Note that the R package `SpatialPack` is used to obtain the noised image.

Visually, for impulse noise, the Median filter is of a better performance than that of the Mean filter, especially for the detail preservation. For comparing results obtained with different windows, the best visual result is achieved by 3×3. The conclusion can also be inferred from Table 5.3. Note that the MSE between the noised and the original images are 546.663.

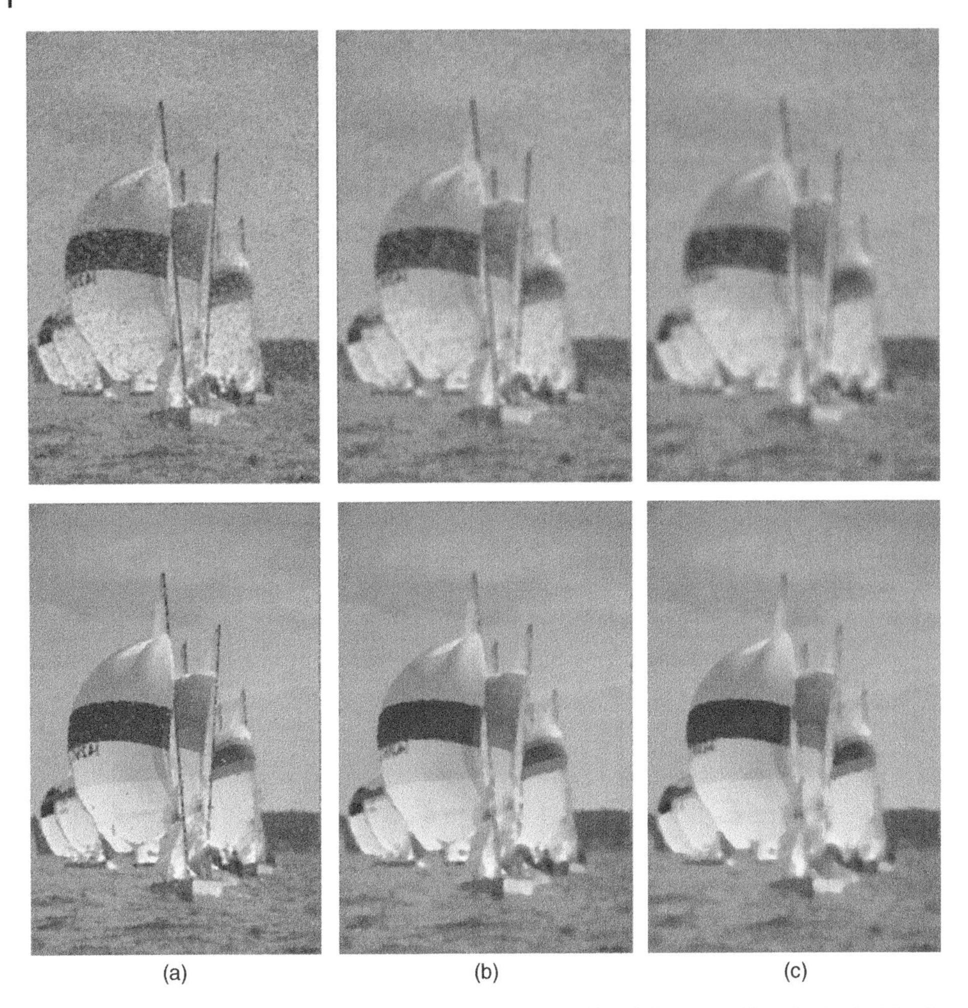

(a) (b) (c)

Figure 5.5 Denoised images with the median filter (bottom) and the mean filter (top) of (a) 3×3 window; (b) 5×5 window; (c) 7×7 window.

Table 5.3 MSE of different windows' size for Impulse noise.

Window size	Mean filter	Median filter
3×3	180.5338	89.8669
5×5	229.9474	165.3348
7×7	279.4417	218.7929

5.1.3 Lee Filter

Under the assumption of local stationary for noise, some spatial filters have been proposed for noise reduction (Lee, 1980, Kuan et al., 1985, Lee et al., 1994, Lopes et al., 1990). Among them, the Lee filter (Lee, 1980) is a simple and effective filtering method for noise reduction. Intuitively, with the local stationary assumption, local first- and second-order moments are computed to estimate the underlying signal. Also, the computation method used for estimation depends on the assumption of the noise model (e.g. additive model or multiplicative model). In Lee (1980), the estimation methods are given in detail for these two noise models.

1. **Additive Model**
   For the additive model, the observed signal is defined as:

$$y = x + \epsilon \tag{5.1}$$

   where y is the observed signal, x is the underlying signal and ϵ is postulated as the white additive noise that is independent of x. This means that $\mu_\epsilon = 0$ and $\mu_{\epsilon_i \epsilon_j} = \sigma \delta_{i,j}$. Here, $\sigma = \sigma_\epsilon$ is the standard variance of the noise, $\delta_{i,j}$ is the Kronecker function and $\mu_.$ is the expectation operator.
   Thus, the strength of the noise is implied by σ_ϵ. Note that, for a good estimation of underlying signal strength with the Lee filter, the value of σ_ϵ should be given or estimated in advance. So, using σ_ϵ, the underlying value corrupted with white additive noise is estimated as

$$\hat{x}_k = \mu_{x_k} + \kappa_k (y_k - \mu_{x_k}) \tag{5.2}$$

   where $\kappa_k = 1 - \frac{\sigma_\epsilon^2}{\sigma_{y_k}^2}$. Note that, for a better estimation of local underlying values, a local sliding window is used in the Lee filter and μ_{x_k} is calculated with μ_{y_k} in the local window. Please refer to Lee (1980) for more details about the inference of Eq. (5.2).
2. **Multiplicative Model**
   For the multiplicative model that is widely used for SAR image modeling, the observed signal (intensity SAR data) is defined as:

$$y = x \times s \tag{5.3}$$

   where y is the observed signal, x is the underlying signal, and s is postulated as the speckle obeying Gamma distribution. Here, $y = f(x, s)$ and we use the first-order Taylor series expansion of y about (μ_x, μ_s). The variable y can be approximated with

$$\begin{aligned} y &= f(\mu_x, \mu_s) + \frac{\partial f(x,s)}{\partial x}|_{(\mu_x,\mu_s)}(x - \mu_x) + \frac{\partial f(x,s)}{\partial s}|_{(\mu_x,\mu_s)}(s - \mu_s) \\ &= \mu_x s + x\mu_x - \mu_x \mu_s \end{aligned}$$

   where μ_x and μ_s are the mean values of x and s. Thus, using the minimum mean square error principle, the underlying signal value is estimated as

$$\hat{x} = \frac{\mu_y}{\mu_s} + \kappa \left(y - \mu_y\right) \tag{5.4}$$

   where

$$\begin{aligned} \kappa &= \frac{C_y^2 - C_s^2}{\mu_s (C_y^2 + C_s^4)} \\ C_y &= \frac{\mu_y}{\sigma_y} \\ C_s &= \frac{\mu_s}{\sigma_s} \end{aligned} \tag{5.5}$$

(a)

(b)

Figure 5.6 Denoised images with the Lee additive filter of (a) 3 × 3 window; (b) 5 × 5 window; (c) 7 × 7 window.

Note that, similar to that of the Lee filter for additive noise, σ_y and μ_y are computed in a local sliding window. For an intensity SAR image, $\mu_s = 1$ and $\sigma_s^2 = 1/L$. While they are $\mu_s = \frac{\Gamma(L+0.5)}{\Gamma(L)L^{1/2}}$ and $\sigma_s^2 = 1 - \mu_s^2$ for an amplitude SAR image. Please refer to Baraldi and Parmiggiani (1995) for more details about the inference of Eq. (5.4).

Here, using the optical image and the corresponding noisy image shown in Figure 5.1, the results obtained via the Lee filter for additive noise are given in Figure 5.6. The corresponding R codes are available in file `Code/R/Chapter6/614.R` from www.wiley.com/go/frery/sarimageanalysis. Moreover, MSE is also calculated and shown in Table 5.4.

The main R codes are as follows:

```
xy <- pixel.grid(noiseImg)
## Lee filtering for additive noise
patches <-
extract_patches(noiseImg,
xy$x,
xy$y,
windowsize,
windowsize,
boundary_conditions = 3L)
denoiseImg <- list()
for (i in 1:length(patches)) {
        meVal <- mean(unlist(patches[i]))
        varVal <- var(unlist(patches[i]))
        if (varVal > noisyVar) {
                kappa <- (varVal - noisyVar) / varVal
                meVal <-
                meVal + kappa * (noiseImg[xy$x[i] + 1, xy$y[i] + 1] - meVal)
        }
        denoiseImg <- c(denoiseImg, meVal)
}
```

Table 5.4 MSE of different windows' size for Gaussian additive noise.

Window size	Lee filter
3×3	34.15437
5×5	31.52959
7×7	33.81313

Note that, the function of `pixel.grid()` is to form a mesh grid of all axes. For the shift caused by the function of `extract_patches()`, the mesh grid should be modified for its second application.

By careful comparison, we can see that the Lee filter achieves the best performance. The main reason is the use of the local adaptive smoothing strategy in the Lee filter. This can also be obtained from MSE (shown in Tables 5.1, 5.2, and 5.4), where the Lee filter with different windows achieves the lowest value of MSE. Besides, using an adaptive smoothing strategy, the Lee filter with a medium window (e.g. 5×5) is of the lowest value of MSE. This is because a smaller window gives less smoothing of noise, while a larger window causes the smearing of the details.

For the multiplicative model, a portion of a real SAR image is shown in Figure 5.7 and processed with the filters mentioned in this section, namely, the Mean filter, the Median filter, and the Lee filter. Note that the Lee filter designed for the multiplicative model is used here for comparison. The visual results are shown in Figure 5.8. The corresponding `R` codes are available in file `Code/R/Chapter6/615.R` from www.wiley.com/go/frery/sarimageanalysis.

Figure 5.7 A portion of SAR image shown in Figure 2.1.

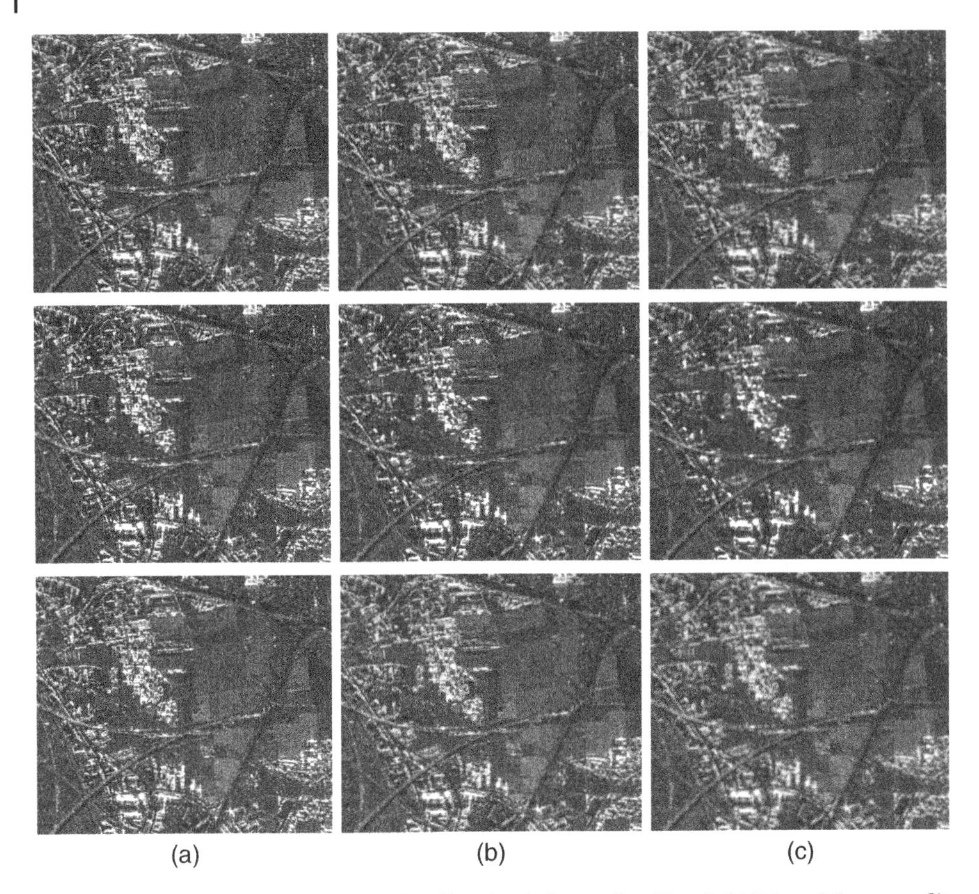

Figure 5.8 Denoised images with the Lee filter (top), the median filter (middle), and the mean filter (bottom) of (a) 3 × 3 window; (b) 5 × 5 window; (c) 7 × 7 window.

The main R codes of the Lee filter designed for the multiplicative noise analysis are as follows:

```
xy <- pixel.grid(as.cimg(noisyImg))
## Lee filter for multiplicative noise
patches <-
extract_patches(as.cimg(noisyImg),
xy$x,
xy$y,
windowsize,
windowsize,
boundary_conditions = 3L)
noisyCV <- 0
if (isIntensity) {
        noisyCV <- 1 / nLooks
        mu_noisy <- 1
} else{
        noisyCV <- 0.273 / nLooks
        mu_noisy <- gamma(nLooks + 0.5) / (gamma(nLooks) * sqrt(nLooks))
}
denoiseImg <- list()
# browser()
for (i in 1:length(patches)) {
        tdata = unlist(patches[i])
        meVal <- mean(tdata)
        varVal <- var(tdata)
        localCV <- varVal / (meVal * meVal)
```

```
        if (localCV > noisyCV) {
                kappa <- (localCV - noisyCV) / (mu_noisy * (localCV
                        + noisyCV ^ 2))
                meVal <-
                meVal / mu_noisy + kappa * (noisyImg[xy$x[i] + 1, xy$y[i] + 1]
                        - meVal)
        }
        denoiseImg <- c(denoiseImg, meVal)
}
```

Visually, a larger window is better for noise reduction, while a larger window will cause a stronger blur of the details. The edge information is better preserved than the Mean filter for the Median filter, while the Lee filter achieves the best edge preservation. The same conclusion is obtained for the conservation of point features. The main reason is that the adaptive strategy is used in the Lee filter. Due to the unknown of the actual signal value of the SAR image, the MSE metric can not be used here for numerical analysis. Please refer to Section 6.1 for more metrics used to assess the performance of despeckling filters.

5.2 Advanced Filters: MAP and Nonlocal Means

After discussing the mean, the median, and the Lee despeckling filters, this section focuses on the Nonlocal means (NLM) and the MAP filters. Both despeckling filters are commonly used in SAR imagery, providing excellent results. For each filter, a brief introduction is first addressed, and then, some application for both, simulated and actual SAR data are given.

5.2.1 MAP Filters

MAP – Maximum A Posteriori filters heavily rely on the statistical modeling of the data. These filters assume that the observation Z, the return, is a corrupted version of the underlying unobserved value X, the backscatter. Provided a model for the observation given the truth $Z \mid X$, a MAP filter estimates X by the "best guess" $X_{\text{MAP}} = \arg\max_X \Pr(X \mid Z) = \arg\max_X \Pr(Z \mid X) \Pr(X)$.

This general setup is particularly attractive under the multiplicative model. We know that $Z \mid X = x$ follows a Gamma distribution with shape parameter L and mean x. With this information, and a model for X with positive support, we can build a MAP estimator.

Let us consider the general case in which $Z = XY$, X follows an Reciprocal Gamma distribution, characterized by density (3.11) and Y follows a Gamma distribution with unitary mean and shape parameter L, whose density is (4.9). The observed return is, thus, the deviate from a $\mathcal{G}^0$ distribution. The MAP estimator is

$$
\begin{aligned}
X_{\text{MAP}} &= \arg\max_X f_{Z|X}(Z) f_X(X) \\
&= \arg\max_X \underbrace{\frac{L^L}{X^L \Gamma(L)} Z^{L-1} \exp\left\{-\frac{L}{X} Z\right\}}_{\Gamma(L,X)\text{ model}} \underbrace{\frac{1}{\gamma^\alpha \Gamma(-\alpha)} X^{\alpha-1} \exp\left\{-\frac{\gamma}{X}\right\}}_{\Gamma^{-1}(\alpha,\gamma)\text{ model}} \\
&= \arg\max_X \left[X^{\alpha-1-L} \exp\left\{-\frac{1}{X}(LZ+\gamma)\right\}\right] \\
&= \arg\max_X \left[(\alpha - 1 - L)\log X - \frac{1}{X}(LZ+\gamma)\right],
\end{aligned}
\tag{5.6}
$$

which becomes

$$X_{\text{MAP}} = \frac{LZ + \gamma}{L + 1 - \alpha}. \tag{5.7}$$

Notice that (5.7) is a linear correction of the return Z. As such, it is unable to reduce speckle.

Since we do not know the local values of α, L, and γ, we form our MAP estimator with Z as

$$\widehat{X}_{\text{MAP}} = \frac{\widehat{L}Z + \widehat{\gamma}}{\widehat{L} + 1 - \widehat{\alpha}}. \tag{5.8}$$

Eq. (5.8) is still a linear correction of the return, but now incorporates local information through the estimators $\widehat{\alpha}, \widehat{\gamma}$ and, possibly, $\widehat{L}$ (this last may be the same for the whole image.) Remind that these are the estimators of $\alpha, \gamma, L \in \mathbb{R}_- \times \mathbb{R}_+^2$, the parameter that indexes the $\mathcal{G}^0$ distribution. Other assumptions require different estimators.

MAP filters are conceptually attractive, because they aim at estimating an unobserved true value under any suitable model. The user can choose which true value she is looking for. In the above derivation we sought for the backscatter, but the MAP filter can be designed with any purpose in mind. Moreover, once we have arrived at, e.g. (5.8), we may decide to use maximum likelihood, moments, robust, or any combination of any kind of estimator.

Moschetti et al. (2006) also derive the MAP estimator for the backscatter under the $\mathcal{K}$ model. Medeiros et al. (2003) obtain and compare several MAP filters for amplitude data. Achim et al. (2006) follow the same path, but assuming a heavy-tailed Rayleigh model.

5.2.2 Nonlocal Means Filter

The filters discussed in Sections 5.1 and 5.2.1 were *local* filters. By local, it is meant that the information gathered for despeckling was taken from the *local* neighborhood of the pixel to denoise. The local mean is the simplest noise reduction filter in image processing. It amounts to applying the same convolution to every pixel with a typically small, e.g. 3×3, $5 \times 5, \cdots$, mask of equal and positive weights. Such a filter, as discussed above, effectively reduces the noise, it is easily implemented, but it also introduces blurring. The Lee filter strongly reduces such blurring effects by filtering in adaptive way to discriminate between homogeneous areas and non-homogeneous areas within the image.

In 2005, a new radical concept for reducing noise in images was published. Buades et al. (2005b, 2010) abandoned the *always used* translation-invariant property, but retained the convolutional paradigm, and proposed a new adaptive approach: the NLM, in which "Nonlocal" refers to the use of large convolutional windows (convolutional masks) to explore areas of the images far from the pixels to be denoised (see Figure 5.9). Such large convolutional mask could be of the size of the whole image although, for practical use (computational cost), its size is much reduced (see below). This class of filters has produced outstanding results in the denoising of natural images (Guo et al., 2016, Liu et al., 2015), Synthetic Aperture Radar – SAR (Deledalle et al., 2009, 2011, Ni and Gao, 2016, Torres et al., 2014) data, and medical images (Santos et al., 2017), among other applications.

In this section, first, the original NLM filter by Buades et al. is summarized and then, a suitable adaptation for SAR images is discussed.

In the original proposal, Buades et al. (2005b) used a Gaussian weighted Euclidean distance between image patches to measure their similarity. Such a measure has good properties with additive white noise, hence well-suited to natural images, but is questionable in the presence

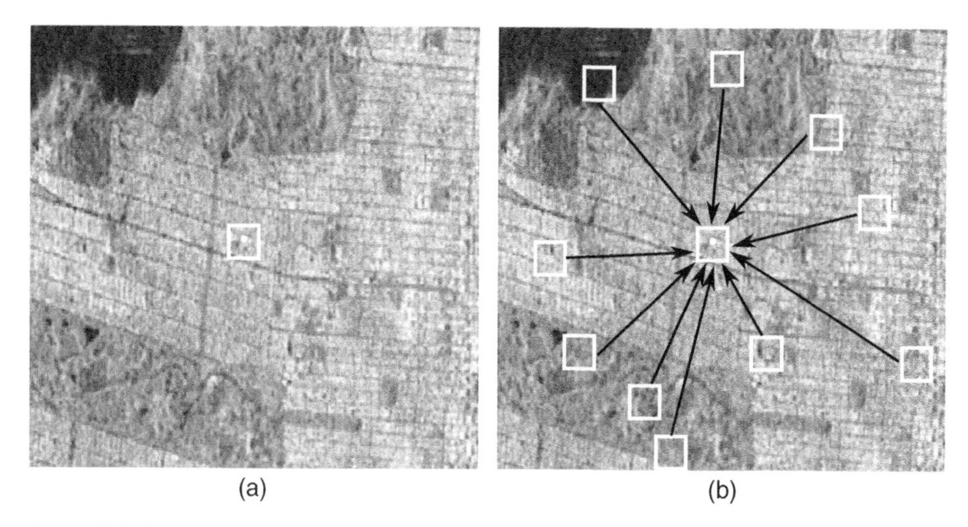

Figure 5.9 Nonlocal means concept: as a radical difference with local filters (a), the information to denoise the central pixel (white dot) may be taken from any area within the image, not matter how far it can be (b).

of other types of contamination. Besides, the mapping between similarities and weights is implicit, leaving little control to the designer.

For an image $z = (z(x))_{x\in\Omega}$ defined in the bounded domain $\Omega \subset \mathbb{R}^2$, and each pixel $z(x)$ is filtered (denoised) as,

$$\text{NLM } z(x) = \frac{1}{N(x)}\sum_{y\in\Omega} w(x,y)z(y), \tag{5.9}$$

where, $N(x)$ is a normalizing factor to assure that the mean values of original pixels are preserved,

$$N(x) = \sum_{y\in\Omega} w(x,y), \tag{5.10}$$

and $w(\cdot)$ is a weight function.

That is, from this NLM formulation, each pixel to be denoised is updated as a weighted sum of pixels taken from the whole bounded domain Ω. However, soon it was realized that it is not necessary to work with all the pixels but just with a suitable semi-local neigborhood, $\Delta(x)$, large enough (usually of size 21×21). This notably reduces the computational cost[1].

Furthermore, the recommended NLM formulation is defined as

$$\text{NLM } \mathbf{z}(x) = \frac{1}{N(x)}\sum_{y\in\Delta(x)} \exp -\frac{\|\mathbf{z}(x) - \mathbf{z}(y)\|^2}{h^2}\ \mathbf{z}(y), \tag{5.11}$$

where $\|\cdot\|$ is the usual L_2-norm, h is a *decay* filter parameter that controls the decay of the exponential function (the decay of the pondering weights as a function of the Euclidean distances), and the normalizing function $N(\cdot)$ is defined as

$$N(x) = \sum_{y\in\Delta(x)} \exp -\frac{\|\mathbf{z}(x) - \mathbf{z}(y)\|^2}{h^2}. \tag{5.12}$$

1 NLM filters show excellent performances but, unfortunately, their computational costs are high.

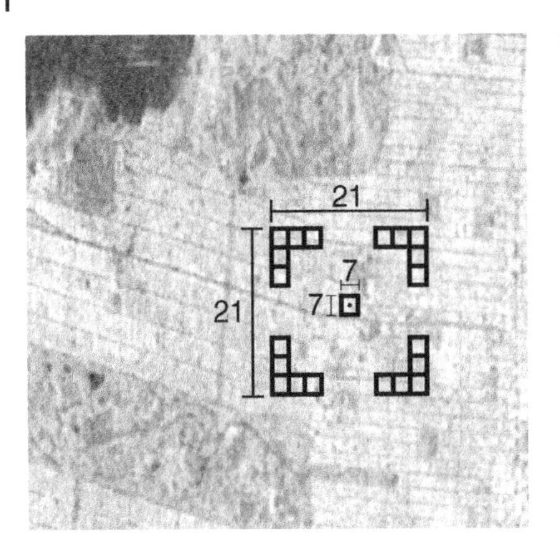

Figure 5.10 Local patch centered at pixel to be denoised is of small size, usually 7 × 7. The semi-local patch is centered at the local patch and its recommended size is of 21 × 21.

In this vectorized formulation, the centered pixel to be denoised is taken in reordered lexicographically to ensemble a n-dimensional vector, $\mathbf{z}(x) := (\mathbf{z}(x_k), x_k \in B(x)) \in \mathbb{R}^n$, where $B(x)$ is a patch centered at pixel x, and it has a size of $\sqrt{n} \times \sqrt{n}$. This patch is known as the local neighorhood or small patch, as a difference with the large patch $\Delta(x)$, denoted above as the semi-local patch (see Figure 5.10). The small patch is usually of size 7 × 7 which allows to capture the local properties (mean values, variance, texture) and also the local features (edges, borders, strong scatterers for the case of SAR images).

From a practical point of view, to apply a NLM filter to an image, the user must provide only,

- size of local patch (by default 7 × 7),
- size of semi-local patch (by default 21 × 21),
- decay parameter value, h (empirically setting in SAR practical cases). However, for natural images corrupted by Gaussian noise, if the standard deviation of the noise, σ is known (or it can be estimated), the recommended value is 10σ (Buades et al., 2005b).

It is relevant to remark that NLM filters deal with edges and fine details in a natural way, that is, no special strategy is necessary (as it is the case for the Lee filter). Additionally, NLM filters preserve the grayscale pixel values of the original image because the updated pixel is estimated from average values of the original grayscale values of the surrounding pixels.

NLM filters established a new paradigm when they first appeared and they perform well and outcome other closely related filters as the bilateral filter (Tomasi and Manduchi, 1998) and other advanced image denoising methods (Mahmoudi and Sapiro, 2005). Additionally, its simple formulation invites to enhance its original capabilities through new enriched statistical formulations, such as Bayesian-based statements and the use of adaptive dictionaries (see, for instance, Kervrann et al., 2007) for denoising natural images or (Zhong et al., 2011) for despeckling SAR images) or by combining look-up tables and variable-size search area to reduce computational cost (see Cozzolino et al., 2014, for a fast implementation specially suited to SAR images). Below, an efficient statistical-based approach for a NLM designed for SAR despeckling is explained.

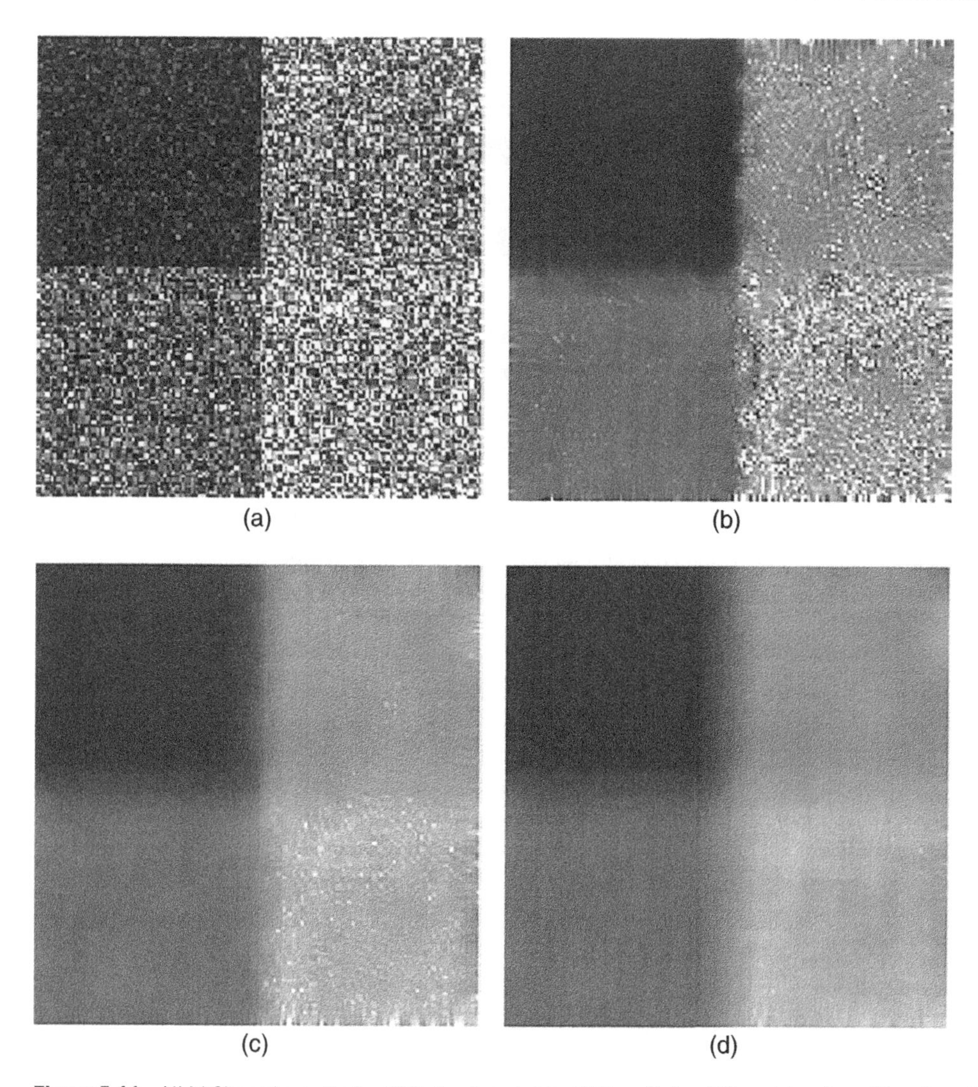

Figure 5.11 NLM filtered results for SAR simulated data (ENL $= 1$) for different h values. The size of patches is the recommended ones: 7×7 for the local patch and 21×21 for the semi-local patch. (a) the simulated data, $h = 2$ (b), $h = 5$ (c), and $h = 10$ (d).

Figure 5.11 shows results for the NLM applied to the simulated data (*the pattern*) for several values of the decay parameter h. As it can be seen, although not originally intended for multiplicative noise, the NLM filter performs reasonable well (ENL $= 1$ is the worst possible case). It is clearly noticed the over-filtered (bad result) result due to a bad setting of the decay parameter ($h = 10$). The user shall find the desired result by a suitable selection of this parameter (for a fixed patches size).

The script to generate the simulated SAR data is located in files

- `Code/R/Chapter6/Build_pattern.R` for R,
- `Code/Matlab/Chapter6/Build_pattern.m` for Matlab.

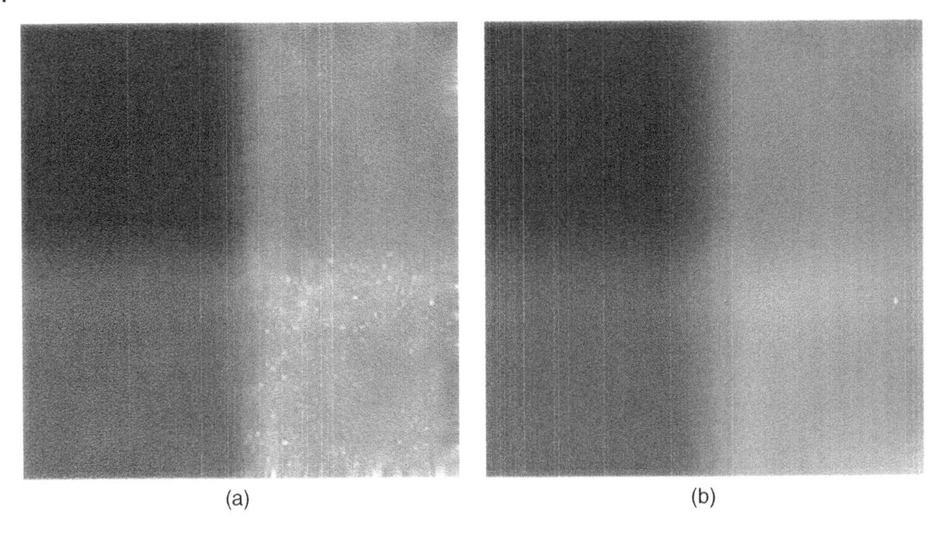

(a) (b)

Figure 5.12 NLM filtered results for SAR simulated data (ENL = 1) for different number of iterations. The size of patches is the recommended ones: 7×7 for the local patch and 21×21 for the semi-local patch. (a) the result for $h = 5$ and one iteration, and (b) the result for 2 iterations.

Although NLM filters are very computationally demanding, as most filters, they can be applied iteratively, that is, the obtained result is filtered again (with the same filter configuration) to improve the previous result. Figure 5.12 shows the result after two iterations for the simulated SAR data (ENL = 1) for one of the cases already shown in Figure 5.11, and $h = 5$. As it can be seen, the speckle content has been notably reduced, but the image is more blurred.

The script to get these results is available in files

- `Code/R/Chapter6/NLM_Script_Experiments.R` for `R`,
- `Code/Matlab/Chapter6/NLM_Script_Experiments.m` for Matlab,

both from www.wiley.com/go/frery/sarimageanalysis.

Matlab provides the method 'imnlmfilt', which is an efficient (compiled version) implementation of the NLM filter (Buades et al., 2005a, b), which is very fast. The NLM filter coded in Matlab, available in the repository www.wiley.com/go/frery/sarimageanalysis, is not a compiled version so, although not optimized in terms of speed, it can be read, easily understood, and also modified to suit user's preferences.

Figure 5.13 shows the result for the NLM applied to actual SAR data. Once again, although not originally intended for multiplicative noise, the NLM filter performs reasonable well, preserving fine details and notably reducing speckle. However, some details are lost, as the small bright pixels on the ocean.

The script to get this result for the actual SAR data is the same file mentioned above,

- `Code/R/Chapter6/NLM_Script_Experiments.R` for `R`,
- `Code/Matlab/Chapter6/NLM_Script_Experiments.m` for Matlab.

The user has the possibility of selecting the data to filter (the simulated SAR data or the actual SAR data).

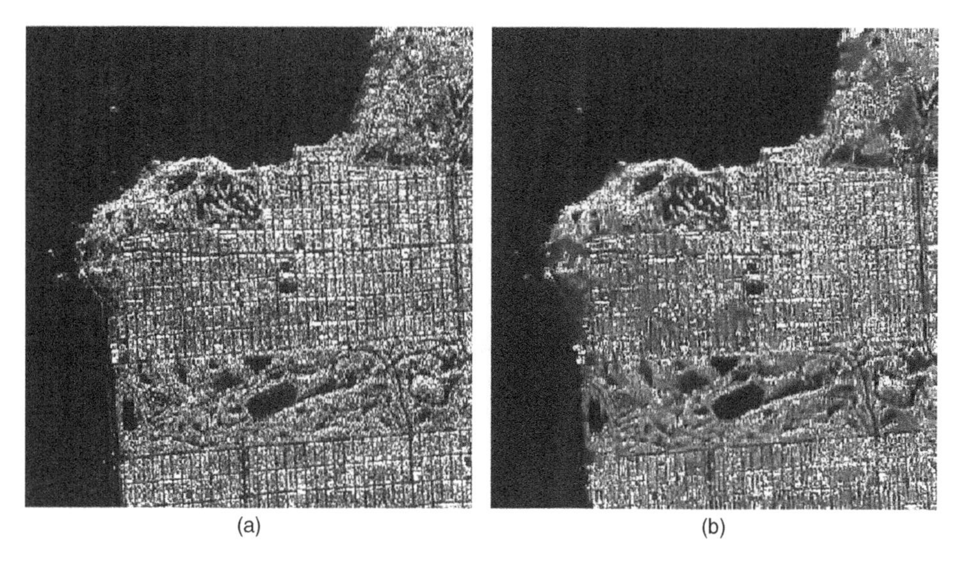

Figure 5.13 NLM filtered results for the San Francisco Bay SAR data with $h = 0.1$ ((a) is the original image and (b) image is the filtered image). The size of patches is the recommended ones: 7×7 for the local patch and 21×21 for the semi-local patch.

5.2.3 Statistical NLM Filters

As detailed above, NLM filters deal with patches instead of single pixels. Then, a set of semi-local patches are compared to the local patch that contains the pixel to be denoised and a weighted sum provides the updated pixel. Apart from other possible simple improvements (for adaptive patch sizes, better coding, ...), the similarity criterion used and how to establish a similarity between patches stems as a worth-researching goal.

Nascimento et al. (2010) obtained for SAR data, tests statistics for the null hypothesis that two samples were produced by the same distribution under the multiplicative model. Such test statistics stem from computing stochastic divergences between the samples and accounting for the peculiarities of these data. This first work handled univariate (intensity) SAR data, and later Frery et al. (2014) extended this approach to the complex-valued multivariate case of PolSAR (Polarimetric SAR) images. In both cases, the approach used h-ϕ divergences to obtain test statistics (Pardo et al., 1995, 1997). Such Information-Theoretic tools allow obtaining an arbitrary number of test statistics with the same asymptotic distribution.

Other works Torres et al. (2014), Deledalle et al. (2009), Wu et al. (2016), and Vitale et al. (2019) discuss ways of computing similarities in the presence of speckle. These works designed different similarities to compute the weights of NLM filters.

Figure 5.14 illustrates the idea of statistical NLM to estimate $9 \times 9 - 1 = 80$ weights for the mask (right grid) that will act around the central pixel (center pixel represented with a white dot). The weight w_1 is an increasing function of p_1, the p-value of the statistical test that assesses the null hypothesis that the samples $\boldsymbol{z}_0 = (z_{0,1}, z_{0,2}, \dots, z_{0,9})$ (see the figure) and $\boldsymbol{z}_1 = (z_{1,1}, z_{1,2}, \dots, z_{1,25})$ were produced by the same probability law $\mathcal{D}(\theta)$, where θ is the parameter that indexes the distribution. This statistical formulation allows using samples of different sizes, i.e. different estimation windows, cf. the dark and light patches, so, providing great flexibility to the design of NLM filters.

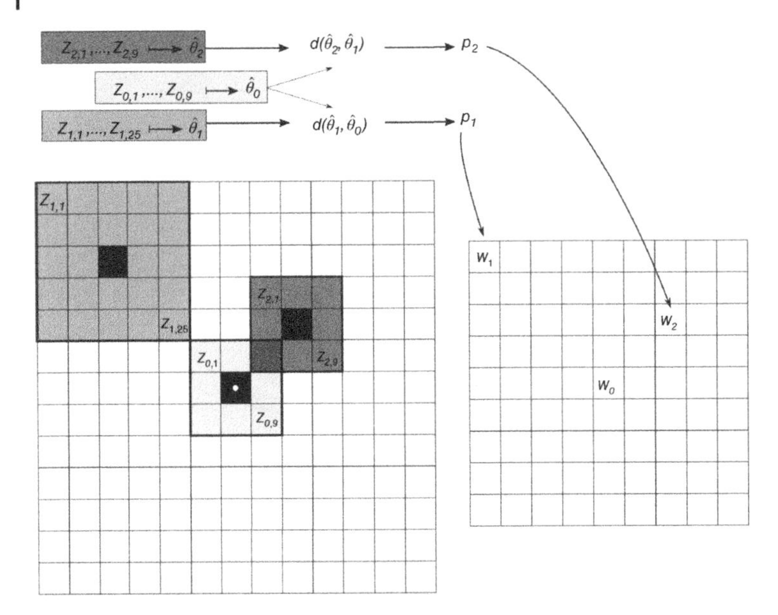

Figure 5.14 Sketch of the statistical NLM approach. The pixel to update is represented with a white dot.

5.2.3.1 Transforming p-Values into Weights

From the sketch shown in Figure 5.14,

- the sample z_0 which surrounds and includes the central pixel is transformed into the estimate $\widehat{\theta}_0$,
- The sample z_1 which surrounds and includes the pixel corresponding to w_1 is transformed into the estimate $\widehat{\theta}_1$.

These estimates are compared by a measure of dissimilarity between the models $d(\widehat{\theta}_0, \widehat{\theta}_1)$. Such a measure, in the context of h-ϕ divergences, is turned into a test statistic:

> The null hypothesis H_0 is that the random vectors $\boldsymbol{Z}_1, \boldsymbol{Z}_2, \dots$ that gave rise to the samples $\boldsymbol{z}_1, \boldsymbol{z}_2, \dots$ obey the same distribution $\mathcal{D}(\theta)$.

Knowing the (possibly asymptotic) distribution of such test statistic, the p_1 and the p-values under H_0 are computed and transformed into the desired weight w_1.

Similarly, the weight w_2 is the result of verifying the hypothesis that the random vectors $\boldsymbol{Z}_0, \boldsymbol{Z}_2$ that produced the samples $\boldsymbol{z}_0$ and $\boldsymbol{z}_2 = (z_{2,1}, z_{2,2}, \dots, z_{2,9})$ follow the same distribution. It is important to remark that the estimates $\widehat{\theta}$ are computed only once for every pixel.

Each p-value may be used directly as a weight w however, as discussed by Torres et al. (2014), such a choice introduces a conceptual distortion. For the samples $\boldsymbol{z}_1$ and $\boldsymbol{z}_2$, when contrasted with the central sample $\boldsymbol{z}_0$ they produced the p-values $p_1 = 0.05$ and $p_2 = 0.93$. In this case, the first weight is significantly smaller that the second one, whereas there is no evidence to reject any of the samples at level $\eta = 0.05$.

Torres et al. (2014) proposed using a piece-wise linear function that maps all values above η to 1, a linear transformation of values between $\eta/2$ and η, and zero below $\eta/2$. Such function

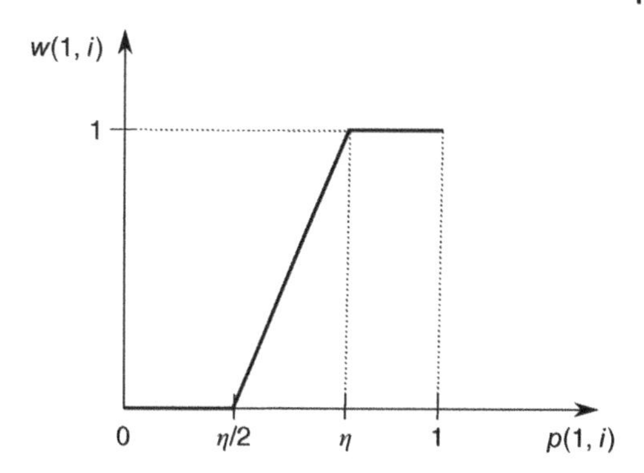

Figure 5.15 Piecewise weight function.

is depicted in Figure 5.15 and is defined as,

$$w(1,i) = \begin{cases} 1 & \text{if } p(1,i) \geq \eta, \\ \frac{2}{\eta}p(1,i) - 1 & \text{if } \frac{\eta}{2} \leq p(1,i) \leq \eta, \\ 0 & \text{otherwise} \end{cases} \tag{5.13}$$

By using this piece-wise weight function, instead of using an accept-reject decision, a soft threshold is employed, avoiding the so-exclusive binary decision criterion.

Returning to the same example discussed above, if the p-values were used instead, the presence of a single sample with excellent match to the central sample would dominate the weights in the mask, forcing other samples that were not rejected by the test, to be practically discarded. For instance, consider the case $p(1; i_1) = 0 : 89$, $p(1; i_2) = 0 : 05$, and all other p-values close to zero. Without the weight function, the nonzero weights would be, approximately, 0.46 and 0.03, so the second observation would have a negligible influence on the filtered value while the first, along with the central value, would dominate the result. Using the aforementioned piecewise function, the three weights would equal 1/3. This notably increases the smoothing effect without compromising the discriminatory ability[2].

Grimson et al. (2015) recommended using values $\eta \in [0.90, 0.99]$ for the significance of the test. The larger the η value, the more similar the patches must be to accept the test hypothesis stated. Additionally, the size of the patches recommended are 3×3 for the local patch and ranging from 5×5 to 11×11 for the semi-local patch. The maximum number of iterations recommended is 3. It is important to note that the recommended size of patches is smaller than for the original NLM filter (7×7 for the local patch, and 21×21 for the large patch).

5.2.4 The Statistical Test

In the original NLM filter, the similarity criterion used for comparing patches was the Euclidean distance. In the statistical NLM filter discussed herein, such a distance criterion is replaced by a statistical divergence measure. The explanation follows in the section is adapted from Grimson et al. (2015).

2 A similar approach to handle the central pixel is used in the original implementation by Buades et al. (2005a), by convolving with a Gaussian kernel the Euclidean distance between two patches. Such a step is omitted in the 'imnlmfilt' method used in Matlab for computational efficiency. For the original NLM filter, it has been also omitted in the R, and in the Matlab codes available in the repository www.wiley.com/go/frery/sarimageanalysis for the same reason.

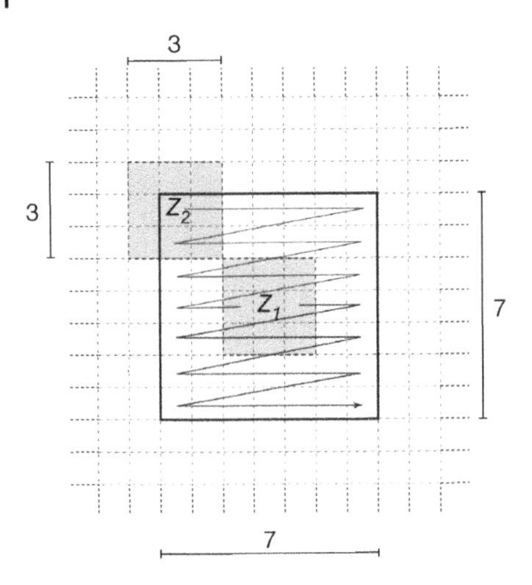

Figure 5.16 Central pixel $\boldsymbol{Z}_1$ and its neighboring $\boldsymbol{Z}_i$, $i = \{2, \dots, 49\}$ with patches 3×3 pixels (local patch) and semi-local patch of size 7×7 pixels.

Denote $\widehat{\theta}_1$ the estimated parameter in the central patch $\boldsymbol{Z}_1$, and $\left(\widehat{\theta}_1, \dots, \widehat{\theta}_n\right)$ the estimate parameters in the remaining areas, where n is the last patch to consider in the neighborhood (for a 7×7 neighborhood and local patches of size 3×3, $n = \{2, \dots, 49\}$ as illustrated in Figure 5.16).

As it was explained in this book, the Gamma law is an excellent model for the return of the SAR signal. The model, with mean λ and number of looks L is characterized by the density,

$$f_Z(z; L, \lambda) = \frac{L^L}{\lambda^L \Gamma(L)} z^{L-1} \exp\left(-Lz/\lambda\right), \tag{5.14}$$

where $z, L, \lambda > 0$. Usually, it is assumed that $L \geq 1$, but, to account for departures from the textureless situation, it can be relaxed since the Gamma law is valid under that assumption.

Therefore, the parameters, $\theta = (L, \lambda)$ for each random sample $\mathbf{Z} = (Z_1, Z_2, \dots, Z_n)$ of independent identically distributed $\Gamma(L, L/\lambda)$ observations can be obtained from the maximum likelihood estimator as,

- $\widehat{\lambda}$, the sample mean: $\widehat{\lambda} = n^{-1} \sum_{i=1}^{n} Z_i$,
- $\widehat{L}$, the number of looks: by solving $\ln \widehat{L} - \psi(\widehat{L}) - \ln \widehat{\lambda} + n^{-1} \sum_{i=1}^{n} \ln Z_i$, which is a single root equation to be solved by any numerical method (Newton-Raphson or similar), with $\widehat{L}_0 = L$, where L is the nominal number of looks of the SAR data (a known data), and ψ is the digamma function.

The statistical test for each pair of estimates for patches Z_i and Z_j can be formulated for any given distance. Following the work by Salicrú et al. (1994), who use stochastic distances, the test for the Kullback-Leibler is given by

$$S_{KL}(\widehat{\theta}_1, \widehat{\theta}_2) = \left(E_r + 1/2\right)^2 \left(\widehat{L}_1 - \widehat{L}_2\right)$$
$$\left(\log\left(\widehat{\lambda}_1/\widehat{\lambda}_2\right) - \log\left(\widehat{L}_1/\widehat{L}_2\right) + \psi(\widehat{L}_1) - \psi(\widehat{L}_2)\right) + \frac{\widehat{L}_1\left(\widehat{\lambda}_2/\widehat{\lambda}_1 - 1\right) + \widehat{L}_2\left(\widehat{\lambda}_1/\widehat{\lambda}_2 - 1\right)}{2}, \tag{5.15}$$

and for the Hellinger stochastic distance by

$$S_H(\widehat{\theta}_1, \widehat{\theta}_2) = 4(2E_r + 1)^2$$

$$\left[1 - \Gamma\left(\frac{\widehat{L}_1 + \widehat{L}_2}{2}\right)\sqrt{\left(\frac{2}{\widehat{L}_1\widehat{\lambda}_2 + \widehat{L}_2\widehat{\lambda}_1}\right)^{\widehat{L}_1+\widehat{L}_2}\frac{(\widehat{L}_1\widehat{\lambda}_2)^{\widehat{L}_1}(\widehat{L}_2\widehat{\lambda}_1)^{\widehat{L}_2}}{\Gamma(\widehat{L}_1)\Gamma(\widehat{L}_2)}}\right], \tag{5.16}$$

where E_r is a scale factor for the distances which relates to the size of the patch where the parameters are estimated (the local patch), given by

$$E_r = (r - 1)/2, \tag{5.17}$$

in which r is the radius of the local patch[3].

Therefore, for a semi-local patch of radius 3 (7×7 mask) as the one shown in Figure 5.16, to perform the hypothesis test between the patches, $\mathbf{Z}_1$ and $\mathbf{Z}_i$, $2 \leq i \leq 49$ the procedure is as follows,

1. For $\mathbf{Z}_1$: Estimate $\widehat{L}_1$ and $\widehat{\lambda}_1$ in the local patch support (3×3 mask in this example),
2. For $\mathbf{Z}_i$: Estimate $\widehat{L}_i$ and $\widehat{\lambda}_i$ in the local patch support (3×3 mask in this example),
3. Get the test statistics (either S_{KL} or S_H from equations (5.15) or (5.16)),
4. Get the p-value as, p-value $= \Pr(\chi^2 > S_{KL})$ (for the case of using the Kullback-Leibler stochastic distance),
5. the null hypothesis $\theta_1 = \theta_i$ can be rejected at significance level η if p-value $\leq \eta$,
6. Compute the weight, $w(1, i)$ from the piece-wise function (equation (5.13)),
7. the central pixel of patch $\mathbf{Z}_1$ is updated as,

$$\widehat{Z}_1 = \frac{1}{\sum_2^n w(1,i)}\frac{1}{n}\sum_2^n w(1,i)\mathbf{Z}_i = \frac{1}{\sum_2^{49} w(1,i)}\frac{1}{49}\sum_2^{49} w(1,i)\mathbf{Z}_i. \tag{5.18}$$

5.3 Implementation Details

As several times pointed out above, NLM filters are computationally demanding, and when compared with other filter techniques such as the Lee filter, not competitive at all at unless an efficient coding is done. Fortunately, it is not complicated to reduce the cost by some actions.

- Parallelized loops: most of the computational cost is due to the multiple comparison performed to test the similarity among patches. This cost is proportional to the size of the semi-local patch and not to the size of the local patch. For instance, for a semi-local patch of size of 21×21, to update the central pixel, $21 \times 21 = 441$ patches must be computed. Obviously, for the whole image, this number must be multiplied by the total number of pixels. That is, for a SAR image of size 1000×1000 pixels (not indeed a large typical SAR image), the total of patches to compute is more than 400 000 000! However, such huge calculus can be done in parallel due to the lack of dependencies among updated pixels. This procedure is easy implemented in most mathematical packages through simple directives. In Matlab, it is enough to replace the 'for' by the 'parfor', and in R by the 'foreach' and 'do Parallel' methods (see https://cran.r-project.org/web/packages/doParallel/doParallel.pdf).

3 For instance, a 3×3 mask has a radius of 2.

- Pre-processing: for the NLM-Statistical filter, the estimates of the Gamma pdf parameters, $\widehat{L}$ and $\widehat{\lambda}$ can be done once and stored in matrices (look-up tables). This avoids to re-compute their values many times when exploring the semi-local patches. It is always recommended to invest enough time in analyzing carefully how much operations can be pre-processed from the inner loops, which would largely compensate the final computational cost.

Sometimes it happens that pre-processing some data can be done easily and quickly. For instance, the estimation of the $\widehat{\lambda}$ parameter is done just by applying a mean filter of mask size equal to the local patch. A similar approach can be done to get all the $n^{-1}\sum_{i=1}^{n}\ln Z_i$ values, but in this case, the mean filter is applied to the $\ln Z_i$.

As an example, the computational cost on an Intel(R) Core(TM) i7-4870HQ CPU 2.5 GHz (16 GB RAM, for the simulated SAR data (the *pattern*), which is of size 150 × 150 pixels, is approximately 5 s for the NLM filter with the recommended parameter setting and approximately 12 s for the NLM-Statistical filter. Computational times for the San Francisco Bay data (400 × 400 pixels) is approximately 35 s for the NLM filter and approximately 90 s the NLM-Statistical filter.

Another issue that is relevant when coding NLM filters is how to manage image borders. This is also necessary for all filters that use masks but, for NLM filters, due to the size of the involved masks, generally, is large (21 × 21). We identify two possible solutions, namely:

- sacrifice the filtering at image borders: this means that the first pixels to be filtered are far enough for borders. For example, for the usual configuration of the NLM filter, the first pixels to be filtered are the ones at position $\lfloor 21/2 \rfloor + \lfloor 7/2 \rfloor = 10 + 3 = 13$, where the operator $\lfloor \cdot \rfloor$ indicates the integer part (the 'floor' command in most programming languages). Therefore, the first pixel to be denoised is the one located at position (13,13). For a small image (for instance, the simulated SAR data used in the examples shown in this chapter), it means that a border of size 13 pixels remains unfiltered.
- Padding the image: it means that the location of the first pixels to be filtered is calculated (for instance, 13), and then the image is expanded outwards and conveniently filled (padding). The new created position can be filled with zeros or other values, for instance by replicating the edge of the image outwards. Then, after filtering, the image is cropped to its original size. This is the strategy employed in the scripts provided.

5.4 Results

Figures 5.17 and 5.18 show some results obtained for the simulated SAR data (ENL = 1) by using the NLM-Statistical filter with two stochastic distances (the Kullback-Leibler and the Hellinger distance respectively). Following the recommendation from the original Authors of this filter (Grimson et al., 2015), small patches are used (not the 7 × 7, 21 × 21 as in the original NLM filter), and more than one single iteration is also applied. The result for the single iteration is already an excellent result and, as noticed, by increasing the number of iterations, homogeneous areas resemble more homogeneous, as expected for a good despeckling filter.

The NLM-Statistical filter gets a better solution for the SAR data than the original NLM filter due to it deals with the particularities of the SAR data,

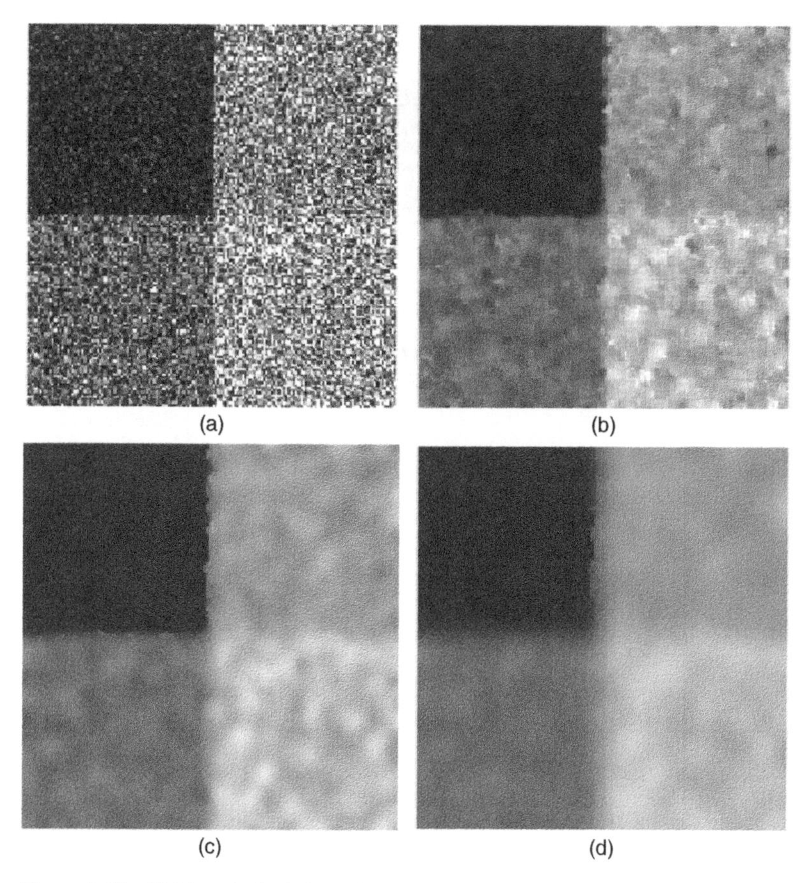

Figure 5.17 NLM-statistical filtered results for SAR simulated data (ENL = 1) for different number of iterations. The size of patches is the recommended ones: 3×3 for the local patch and 7×7 for the semi-local patch. (a) the simulated data, (b) one iteration, (c) two iterations, and five iterations (d). The Kullback-Leibler divergency was used as similarity criterion among patches and $\eta = 0.95$ was set for the significance test.

- it takes into account the expected distribution of the SAR data through the gamma model, which includes the number of looks of the data (a physical value directed related to the SAR data acquisition/processing system),
- it uses a more sophisticated criterion to compare the patches (statistical divergences) instead of the Euclidean distance,
- it applies a well-suited to SAR data statistical hypothesis test.

Figure 5.19 shows the results for the NLM original filter and for the NLM-Statistical filters using the Hellinger stochastic distance and the Kullback-Leibler stochastic distance. These results were already shown in previous figures and they are put together for the sake of better comparing the NLM filters. The results clearly are favurable for the NLM-Statistical filters, and the use of the Kullback-Leibler divergence provides much speckle reduction than the use of the Hellinger divergence although significantly blurring the image.

In Figure 5.20, the estimated ENL, $(\widehat{L})$, values for the simulated SAR data are shown. Although the nominal simulated data was $L = 1$, finally, $\widehat{L}$ ranges from $1/2$ to 3, but, its average value is very close to 1 everywhere, as expected.

Figure 5.18 NLM-statistical filtered results for SAR simulated data (ENL = 1) for different number of iterations. The size of patches is the recommended ones: 3×3 for the local patch and 7×7 for the semi-local patch. (a) the simulated data, (b) one iteration, (c) two iterations, and five iterations (d). The Hellinger divergency was used as similarity criterion among patches and $\eta = 0.95$ was set for the significance test.

For the NLM-Statistical filter, the η parameter (the significance of the test) plays a similar role than the decay parameter h for the original NLM filter, that is, controlling the amount of blurring on the filtered result. Figure 5.21 shows a result with $\eta = 0.1$, which implies that almost all patches surrounding a given central pixel contribute with the same weight to the sum to update the pixel to be denoised. In this case, the NLM filter works as a simple mean filter, thus, providing over-filtered results even for a single iteration.

The script to get these results is available in files

- `Code/R/Chapter6/NLM_Statistical_Script_Experiments.R` for `R`,
- `Code/Matlab/Chapter6/NLM_Statistical_Script_Experiments.m` for Matlab,

both from www.wiley.com/go/frery/sarimageanalysis.

These scripts are also the same used to get the results for the actual SAR data (see below)

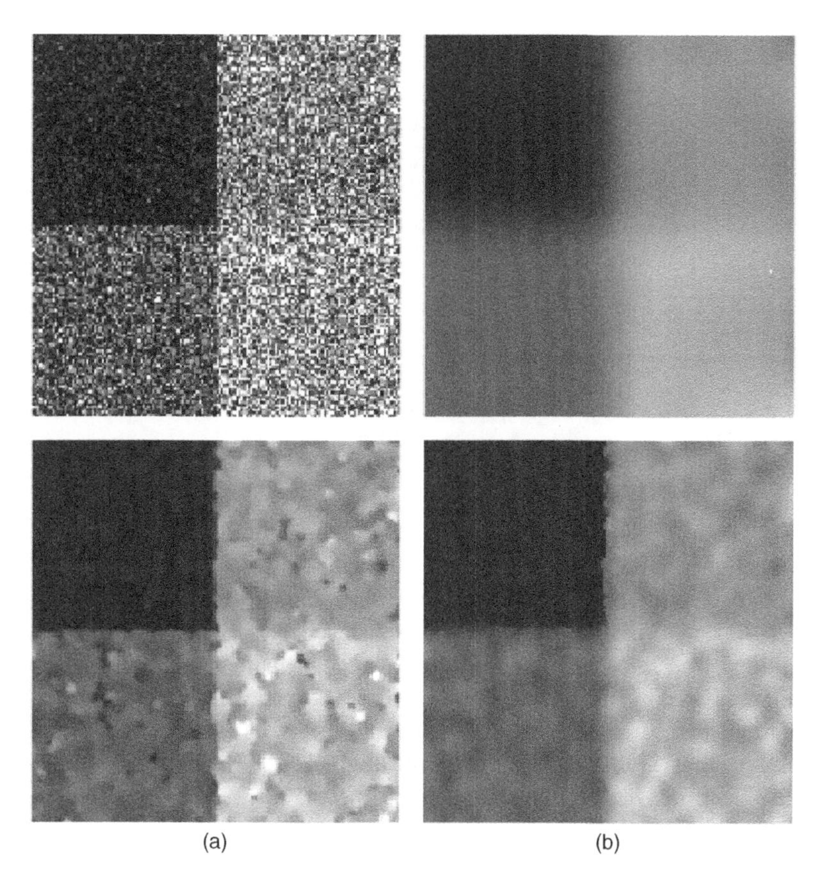

Figure 5.19 NLM filter and NLM-statistical filter comparison for the simulated SAR data (ENL = 1). (a) the simulated data, the NLM result (b), the NLM-Statistical filter with the Hellinger stochastic distance, and the NLM-Statistical filter with the Kullback-Leibler stochastic distance. The setting for the filters are the same used before in previous examples. The number of iterations is two for the three filters.

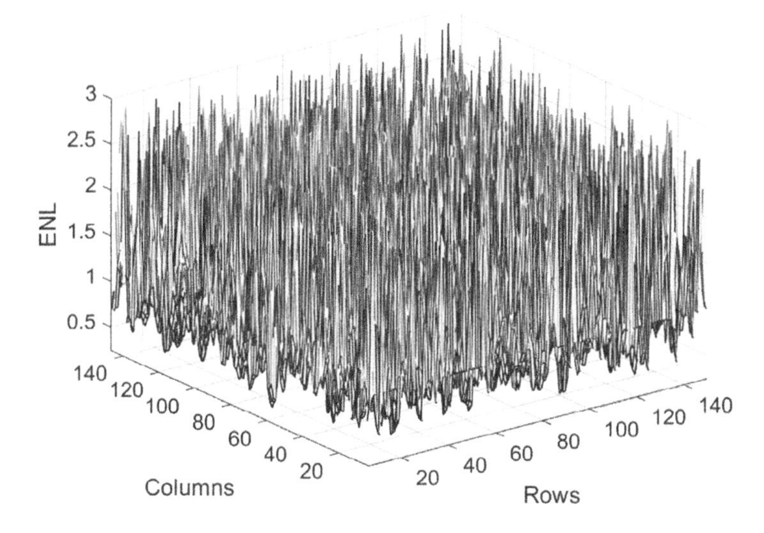

Figure 5.20 Estimated ENL, $(\widehat{L})$, for the simulated SAR data.

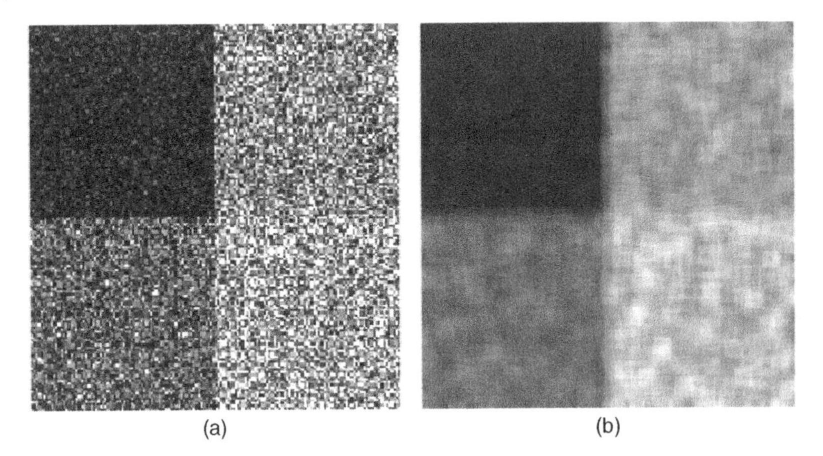

Figure 5.21 (a) Simulated SAR data (ENL = 1) and an over-filtered result for the NLM-Statistical filter for $\eta = 0.2$ (b).

The last result is for the actual SAR data. Figure 5.22 shows the result for the San Francisco Bay 4 Looks data obtained with the NLM-Statistical filter using the Hellinger stochastic distance and $\eta = 0.95$ and a single iteration. The filter performed well, preserving the bright and fine details and making homogeneous areas more homogeneous. Figure 5.23 shows a zoom of the figures (original, filtered by the NLM filter and by the NLM-Statistical filter) to better visualize how both filters performed. The result from the NLM-Statistical filter contains little speckle and preserve more details. The result from the original NLM shows some missing details (the shape big black area has been modified).

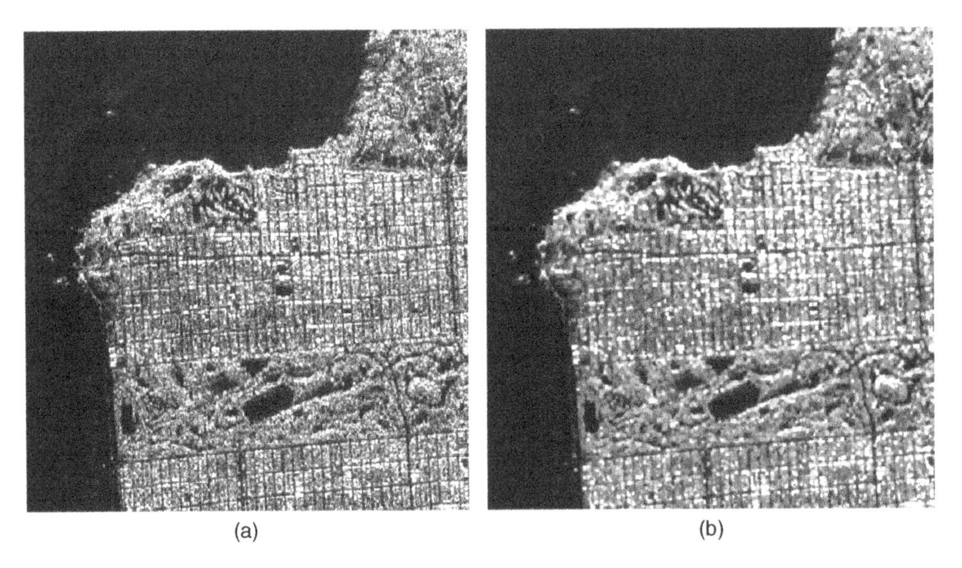

Figure 5.22 NLM-statistical filter result for the San Francisco Bay SAR data with $\eta = 0.9$ ((a) is the original image and (b) image is the filtered image). The size of patches is: 3×3 for the local patch, and 7×7 for the semi-local patch. The number of iterations was set to one.

Figure 5.23 NLM-Statistical filter result for the San Francisco Bay SAR data with $\eta = 0.9$ ((a) is the original image and (b) is the filtered image). The size of patches is: 3×3 for the local patch, and 7×7 for the semi-local patch. The number of iterations was set to one.

5.5 Classification

Speckle content in SAR data makes it difficult to interpret the acquired images. This is also true even for SAR experts. On the other hand, huge amount of SAR data is generated every day and unsupervised analysis is demanded to quickly identify targets in routinely done tasks such as monitoring of urban growing or deforestation, among other common Remote Sensing applications.

Once again, SAR benefits from the plethora of methods developed for natural images over the last years. In this case, we focus on Image Classification and, the required particularities due to SAR data must be conveniently taken into account to suit the standard techniques to SAR images.

In the context of Image Processing, the image classification problem consists of assigning a label to group of pixels based on specific rules. The group of pixels is, therefore, assigned to a class where they shared some common characteristics (color, texture, variance, etc.) The number of classes can be known a priori or be completely unknown.

Classification has been a central subject of research within Machine Learning methods and many techniques, and algorithms are available. Most mathematical packages, such as Matlab, Mathematica, R include many classification methods.

Classification methods are divided into unsupervised methods and supervised methods. The unsupervised methods (automatic classification without the intervention of a user) are desirable, however due to the complexity of SAR data, most methods are either semi-supervised or supervised.

The supervised methods use a set of labeled samples (labeled by an expert) that share some common features. Therefore, the number of classes to classify all data in is 'a priori' known. The initial set of classes is known as the *training set* and the supervised method extracts information from the training set to complete the classification of the whole data space. It is clear that the *quality* of the training set has a direct impact on the final result: a bad initial selection traduces, generally, in a bad final classification result. Unsupervised methods do not require the intervention of a user (expert) and operate by extracting information from data to guess the number of the classes (not 'a priori' known) and then, automatically assigning the data into the classes. Usually, this is done in an iterative way (the number of classes changes as the algorithm converges) following some minimum residual error criteria (see below). Most of the supervised methods in image classification, and so, in SAR image classification, use statistical information from the data to build complex machine learning methods such as support vector machines or random forests.

From above, a supervised classification method maps the input variable, X (objects to classify) to an output variable (Y) through a mapping function $f(\cdot)$,

$$Y = f(X), \tag{5.19}$$

that is learned from the input-data correspondences. Once the mapping function is known, any new variable is classified. However, unsupervised methods work with input data and no input-output data correspondences is known. For the unsupervised methods, the aim is to extract as much as possible, knowledge directly from data without user intervention.

Supervised methods are, in general, methodologically and computationally simpler. Unsupervised methods provide less accurate results due to data is not previously labeled (by user/expert) and the result obtained commonly requires interpretation. Those are the reasons for the extensive use of supervised methods in image processing for tasks such as data classification.

A summary of the most used classification methods for image processing is shown in Table 5.5. From those methods, the regression method (supervised) and the K-means clustering have been selected for the applications shown in this chapter. Additionally, deep learning methods have emerged recently with, in most cases, outstanding performances and it seems that the trend is to replace all standard methods when dealing with massive data classification problems. Deep learning application to SAR data is beyond the objectives of this chapter. For both, deep learning methods (CNN, convolutional neural networks) and the large plethora

Table 5.5 Most used classification methods in image processing.

Unsupervised	Supervised
K-Means clustering	Regression methods
Principal component analysis (PCA)	Support vector machines (SVM)
—	Random forests
—	Neural networks

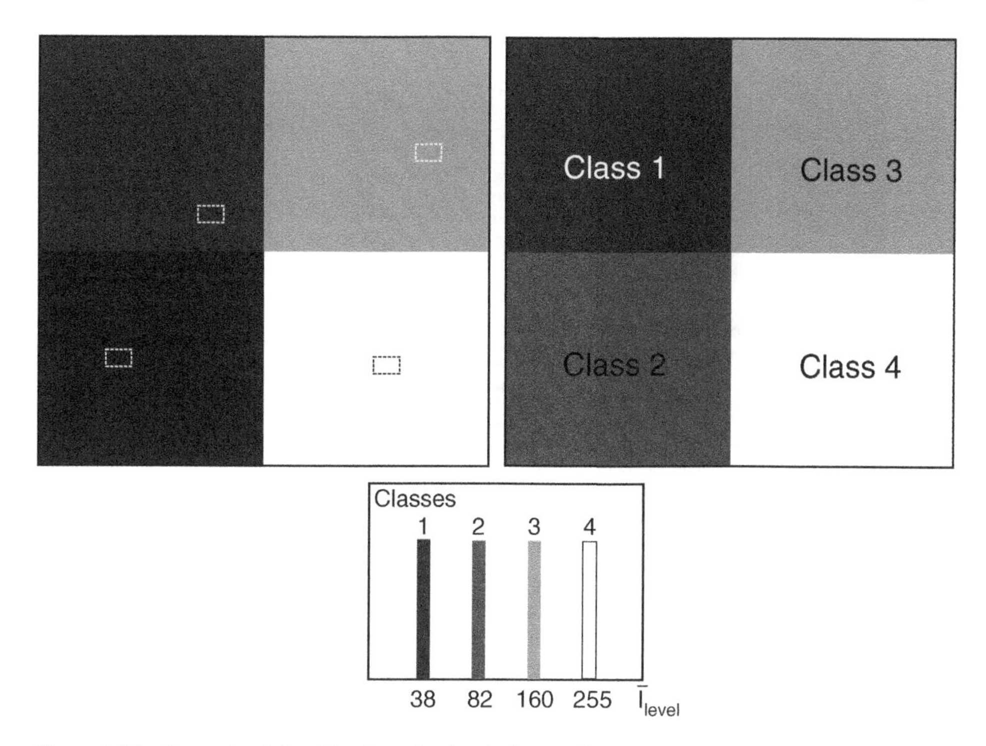

Figure 5.24 Supervised classification of a simple image. The small regions within each class in dashed line are the training data selected by the user.

of machine learning classification methods (supervised, unsupervised, semi-supervised) the reader is referred to the vast bibliography available (see, for instance, Hastie et al., 2017, Goodfellow et al., 2016, and published papers from remote sensing journals).

Figure 5.24 illustrates with a simple example the problem of classification. The image on the left is just a pattern with four disjoint classes. The task is to classify the image. By using a supervised method, the user selects some training data for each class (the number of classes is known), and also specify some criterion to classify similar pixels (the *features*). Then, the algorithm chosen solves the problem and each class is correctly identified and marked (top right figure). In this case, the selected feature was just the average grayscale value of the pixels of each training set, which are well separated in their gray level values (38 for class 1, 82 for class 2, 160 for class 3, and 255 for class 4).

However, the problem can be more difficult to solve if the data is corrupted with noise, as illustrated in Figure 5.25. The pattern now shows speckle ($L = 1$), and sets are not disjoint as they were before. Correspondingly, its classification is a harder problem ($L = 1$ is the worst case), as shown in Figure 5.27, where most of the pixels are not properly classified (top right figure) by means of the naïve classifier (minimum distance criterion to the nearest representative of each class). However, by using a despeckling filter (the simple median filter in this case), the classification improves remarkably, although misclassified pixels are still visible and edges are largely degraded due to the overlapped among pixels distributions.

The result shown in Figure 5.26 was obtained using Matlab with the script available in file `Code/Matlab/Chapter6/Histograms_pattern.m` from www.wiley.com/go/frery/sarimageanalysis.

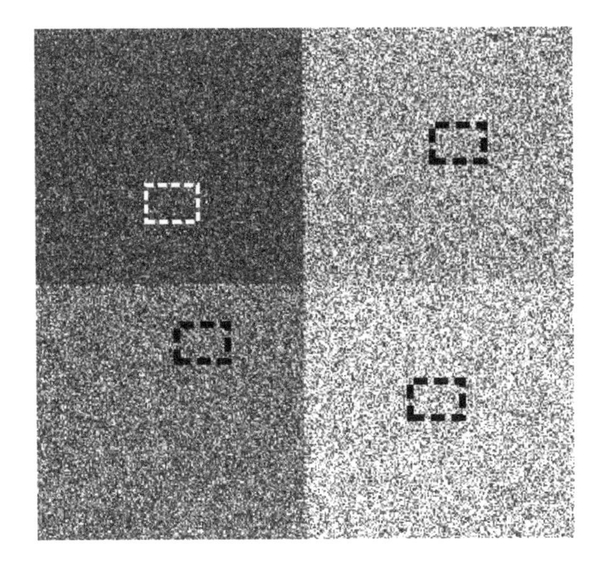

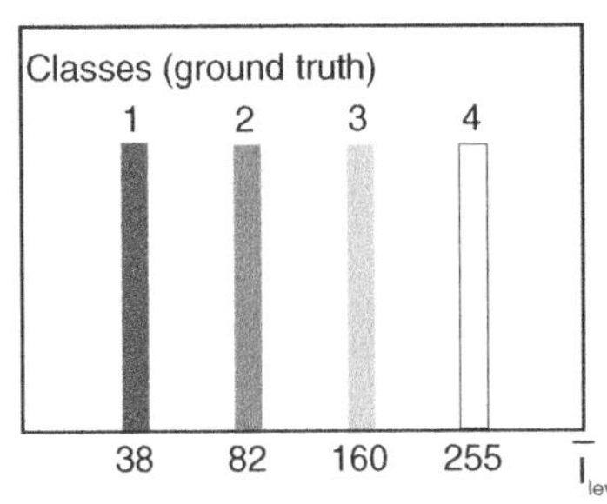

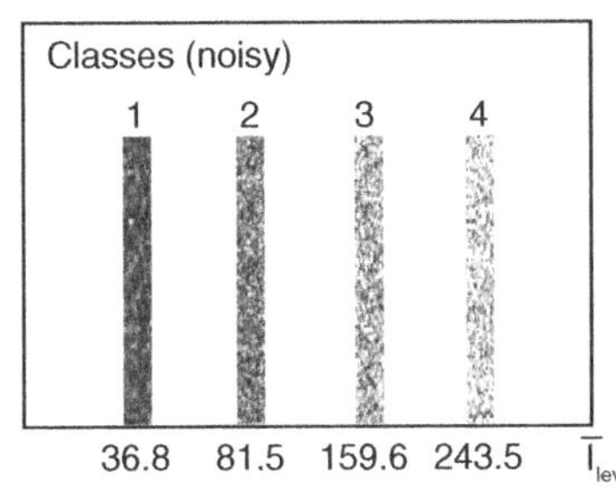

Figure 5.25 Simulated speckled image. The small regions within each class in dashed line are the training data selected by the user.

The result shown in Figure 5.27 was obtained using Matlab with the script available in file `Code/Matlab/Chapter6/Nearest_centroid.m` from www.wiley.com/go/frery/sarimageanalysis.

For actual SAR images, a result for a basic classification method (nearest centroid) is shown in Fig 5.28. The image was filtered by the median filter. Three classes were selected, ocean, forest, and urban area. The result for a low-profile classification method as the one used can be largely improved by other techniques.

Result shown in Figure 5.28 was obtained using Matlab with the script available in file `Code/Matlab/Chapter6/Nearest_centroid.m` from www.wiley.com/go/frery/sarimageanalysis.

It is interesting to note that, if instead of dealing with SAR data (single polarization mode only available), one deals with PolSAR (Polarimetric SAR) data, where more than a polarization mode is used, classification result improves considerably, as it is the case shown in Fig 5.29, from Gomez et al. (2017a). Five classes were selected in this example, ocean, sand, grass, forest and urban.

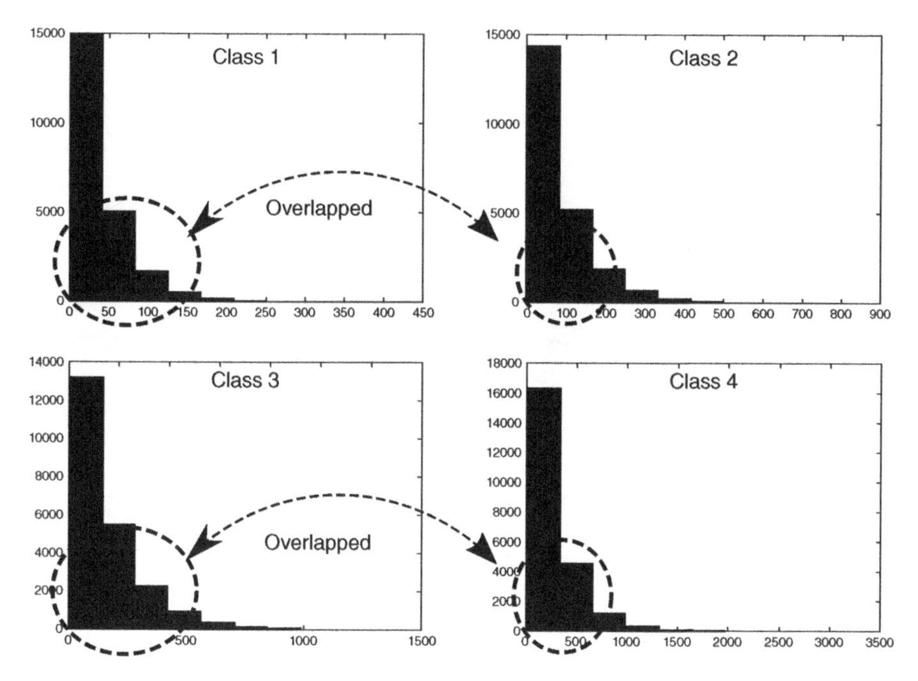

Figure 5.26 Histograms of the four speckled areas (overlapped is noticeable).

For SAR data, the expected results are not as good as for the case of dealing with PolSAR data due to less information is available but, as it is explained in Section 5.6, by using better classification techniques, the result for SAR data will significantly improve.

5.5.1 The Image Space of the SAR Data

Either using a supervised or an unsupervised method, data is the primary source for any classification method.

SAR data (as most image data) comes in a 2D-array of elements or a matrix, A of dimensions $m \times n$, where m is the number of rows and n is the number of columns (see Figure 5.30). Each $a_{i,j}$ position ($i \in \{1, m\}, j \in \{1, n\}$), stores the backscattered value corresponding to the energy reflected from the Earth's illuminated area by the synthetic antenna. Each SAR data, as explained in Chapter 3, has amplitude (or intensity) and phase information, so, it is a complex number.

It is important to remark, as it has been done several times in this book, that one thing is the codification of the image for its visual representation and another thing is the true values of the image (SAR data). For example, in the case illustrated in Figure 5.30, the SAR data has been quantized to 8 bits, so, a 0 value is for a black pixel and a 255 value is for the brightest value (full white color). However, to apply an algorithm to numerically processing the SAR data for a task such as filtering or **classifying**, true backscatter values must be considered. For the single SAR image shown, the true values are shown (a reduced set of them)

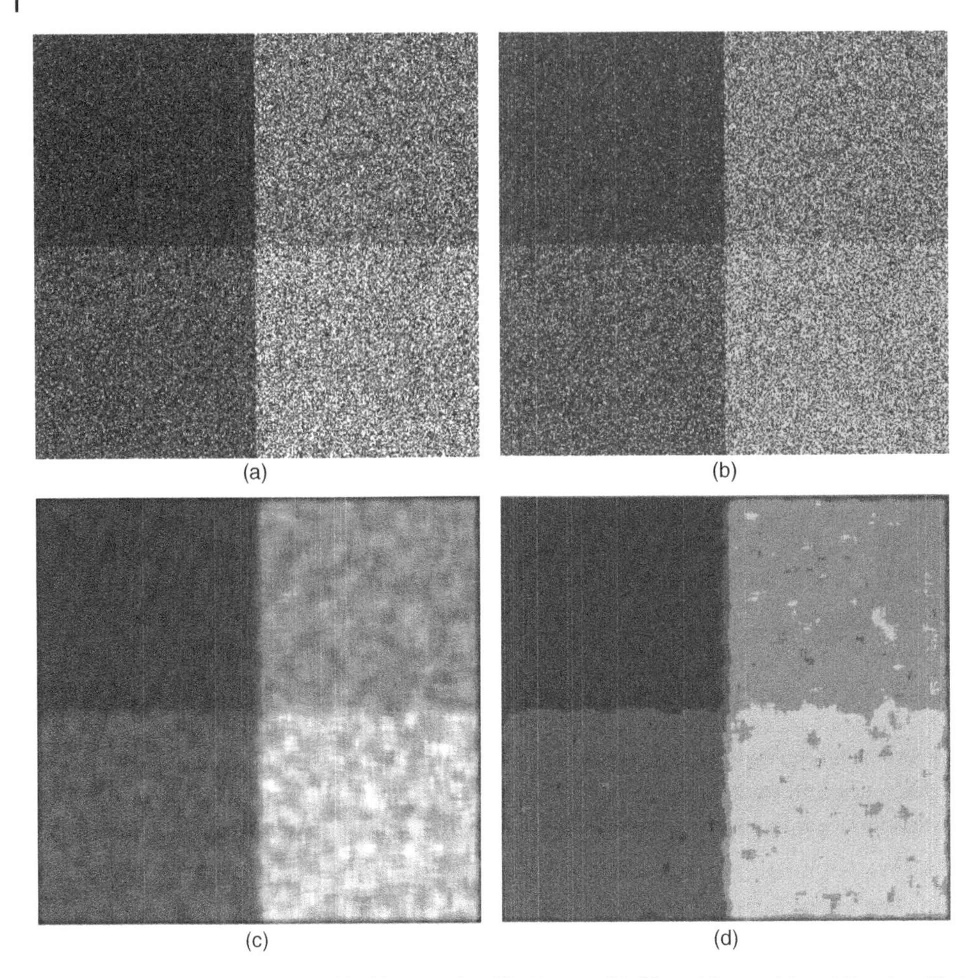

Figure 5.27 (a) Simulated speckled image, classified image (b), filtered image (c) and the classified image(d).

Figure 5.28 (a) SAR image (a selected area of San Francisco Bay) and classified image (b). The image has been filtered by a median filter before applying the *naive* classification method.

Figure 5.29 (a) PolSAR image (a selected area of San Francisco Bay and classified image by an advanced classification technique (b).

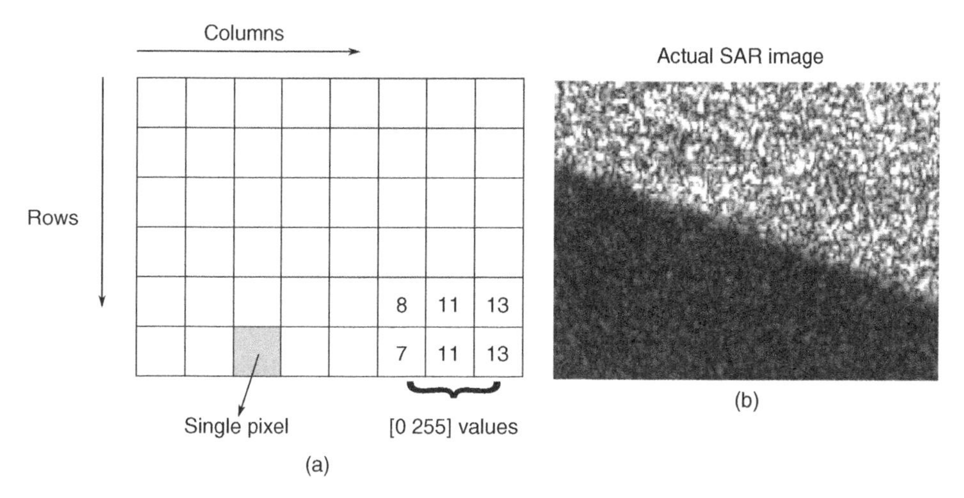

Figure 5.30 (a) Image stored as a 8 bits grayscale matrix (pixel values range from 0 to 255) and, an actual SAR image (b). Values shown in the matrix correspond to true values for the image shown.

in Figure 5.31, and those ones are the values that will be used in all the methods discussed below. Additionally, the SAR data may come in intensity or amplitude mode.

Result shown in Figure 5.30 was obtained using Matlab with the script available in file `Code/Matlab/Chapter6/Example_quantization.m` from www.wiley.com/go/frery/sarimageanalysis.

5.5.2 The Feature Space

Feature is a key concept in classification, and it is defined in most dictionaries as "a typical quality or an important part of something" (Cambridge Dictionary). To classify an object, it is mandatory to select its appropriate set of features to assure a fine result. The better the

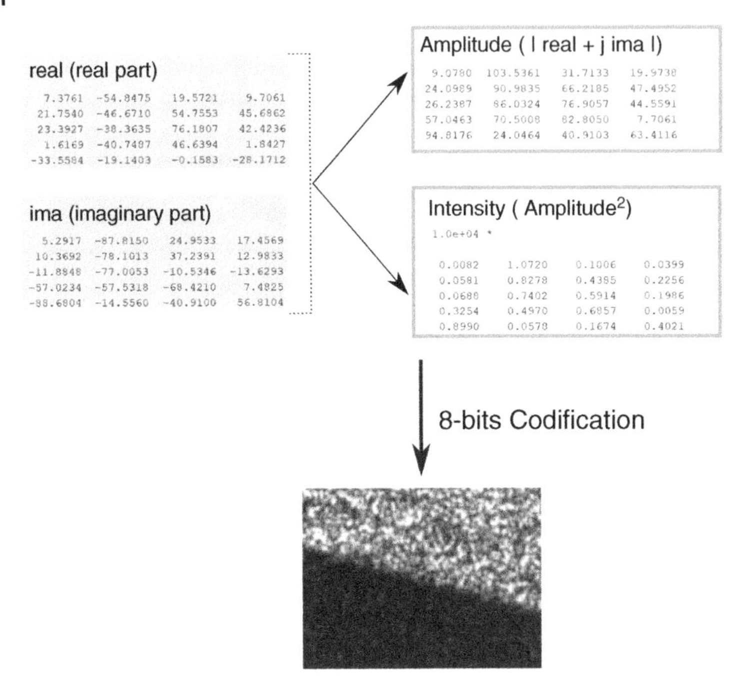

Figure 5.31 From true backscatter values to gray level values. Algorithm must operate with true backscatter values (either in intensity or amplitude mode).

object is described through its features, the better the result will be. Future selection is a large area of research due to it is not a trivial task to identify the best features for an object. As an exercise, the reader can dedicate some time to write down what features unequivocally identify a simple object, for instance, a chair. Soon, many conflicts will appear: a chair has always 4 legs?. If noise and distortions are added to the initial image, the complexity of the classification tasks increases significantly. For single-polarization SAR images, the *natural* feature space is defined by the pixel values (the backscatter value). Hence, the SAR feature space is a vector of one dimension, $\mathbf{v} = a_{ij}$. It is expected that pixels that belong to an object within the image will share similar pixel values and those values will be significantly different from the pixel values for other objects. This is illustrated in Figure 5.32.

Obviously, a more sophisticated election is possible (through for instance mathematical transformation such as wavelets or so), but for the introductory aim of this section, the pixel value will be enough. However, from the pixel values and, avoiding too complex mathematical techniques, many useful object descriptors can be built. For example, specific image texture descriptors (see for instance the Haralicks descriptors, Haralick et al. (1973)) are easily calculated through mathematical packages (GLCM package in R). Haralick's descriptors (28 textural features extracted from each of the gray-tone spatial dependence matrices, or co-ocurrence matrices) will define a large feature space that can complicate the classification process (causing overfitting problems). But, just using a reduced number of them, for instance, 2, it will result in a feature vector as $\mathbf{v} = (v_1, v_2)$. It is interesting to note that Haralick's descriptors, sic. 'are based on statistics which summarize the relative frequency distribution (which describes how often one gray tone will appear in a specified spatial relationship to another gray tone on the image).', from Haralick et al. (1973). Figure 5.33 shows an example of image classification by using some Haralick's textural

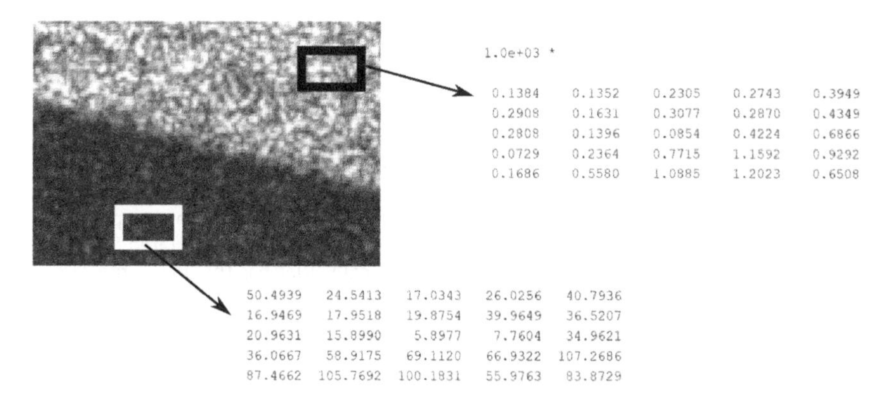

Figure 5.32 Single look SAR image showing two well-different areas (upper and lower). Some pixel values for the two selected training sets are also shown (solid white line rectangles). It is clear that the pixel values lie in enough-separated regions of the feature space.

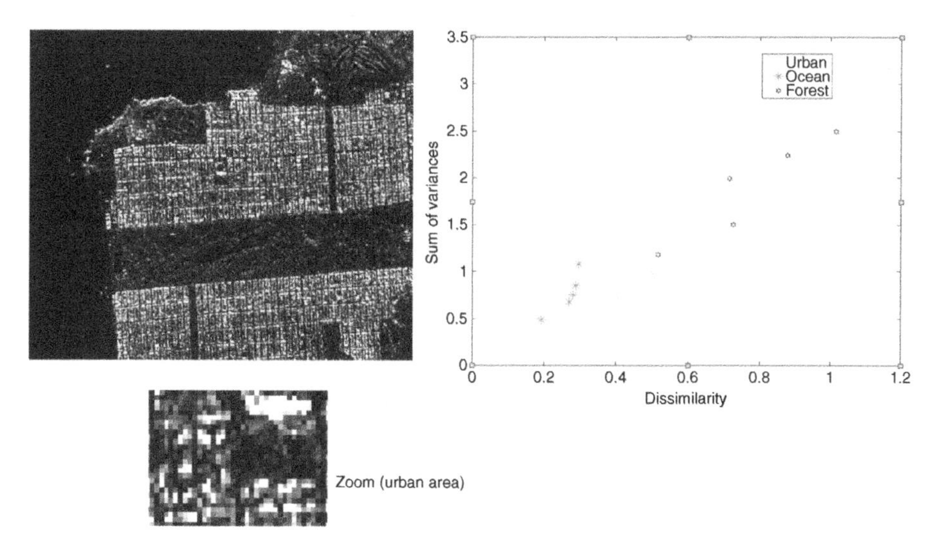

Figure 5.33 Feature space defined by two Haralick's texture descriptors (dissimilarity and sum of variances) for the San Francisco Bay image. The data groups into three sets (clusters): ocean, forest and urban. The zoom for the urban area shows that it is not a disjoint class due to its content of forests (gardens, trees) which is also reflected into the scattering of urban and forest data in the feature space (data for these two clusters overlap). However, ocean data is clearly grouped and separated from the other clusters.

descriptors. Results shown in this figure were obtained using Matlab with the script available in file `Code/Matlab/Chapter6/Example_Haralick_classification.m` from www.wiley.com/go/frery/sarimageanalysis.

5.5.3 Similarity Criterion

Once the feature has been selected (for instance, the pixel value) and noting that objects (targets, areas,...) that belong to an object share similar features, the next step is to select a *similarity criterion* to discriminate between different objects to classify. A natural selection for estimating how far are two features in the feature space is just to calculate their Euclidean

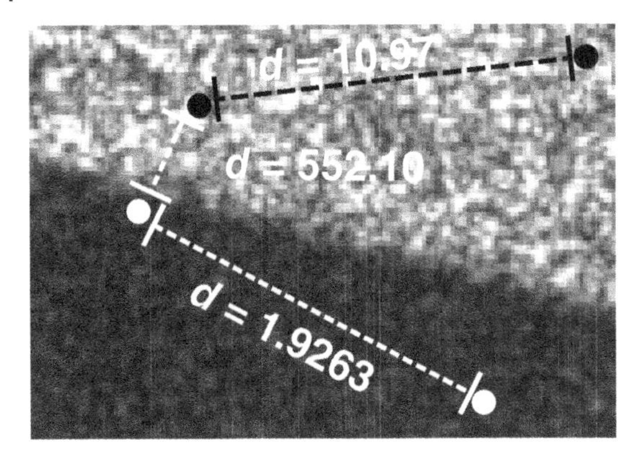

Figure 5.34 Euclidean distances (absolute difference of grayscale pixel values) for pixels belonging to same objects ($d = 10.97$, and $d = 1.9263$) and different objects ($d = 552.10$). It is interesting to note that two pixel may lie close within the image and show a large *euclidean distance* in the feature space. The euclidan distance calculated from the numerical values has the unit of the axis.

distance [4]. Hence, for two features defined by $\mathbf{v}$ and $\mathbf{w}$, the distance is calculated as the module of the vector $\overrightarrow{\mathbf{vw}}$, that is, $d(\mathbf{v}, \mathbf{w}) = ||\overrightarrow{\mathbf{vw}}||$. If the selected features were their grayscale values ($\mathbf{v} = \left(a_{i,j}\right)$, $\mathbf{w} = \left(a_{i',j'}\right)$, where i, j and i' and j' are the coordinates of the two pixels in the SAR image), the distance is calculated simple as the absolute value of numerical difference of the pixel values, $d = |a_{i,j} - a_{i',j'}|$. As expected, see Figure 5.34, for similar regions (target, objects,...) the Euclidean distance in the feature space is low, and it increases with the dissimilarity between pixel values.

If more descriptors are used for the features, the distance is calculated as usual. For instance, if two of Haralick's descriptors were considered, $v = \left(v_1, v_2\right)$, $w = \left(w_1, w_2\right)$, the distance between two features is,

$$d(v, w) = ||\overrightarrow{\mathbf{vw}}|| = \sqrt{(w_1 - v_1)^2 + (w_2 - v_2)^2}. \tag{5.20}$$

5.6 Supervised Image Classification of SAR Data

Two basic image classification methods are explained in this section. The first one used a minimum distance-based criterion (nearest neighbor) and the second a probabilistic approach (maximum likelihood classifier). From those methods, it is not complicated to design more sophisticated techniques and also, to understand most of the existing techniques.

First, a general view for a supervised method is illustrated in Figure 5.35, which summarizes all explained above. As it can be seen, the supervision by an expert (or user) is done in many stages of the process: to select the training sets, to select the classification method, to run the method and check results, and to perform the final definitive evaluation through some specific metrics (assessment). The assessment requires to have ground truth data (reference data) to compare with, which is not a trivial issue when dealing with SAR data. More on assessment is discussed later on.

It is important to remark that SAR data can be despeckled (at least its content of speckle significantly reduced) from the starting of the classification process illustrated in Figure 5.35. Another possibility is not to despeckled the SAR data when training the method and only despeckled the image to be classified. One must be aware of the loss of information when applying a despeckling filter, but also, one must be aware of the difficulty of the classification

4 Other distance can be used, see for instance Naranjo-Torres et al. (2017) for the use of stochastic distances.

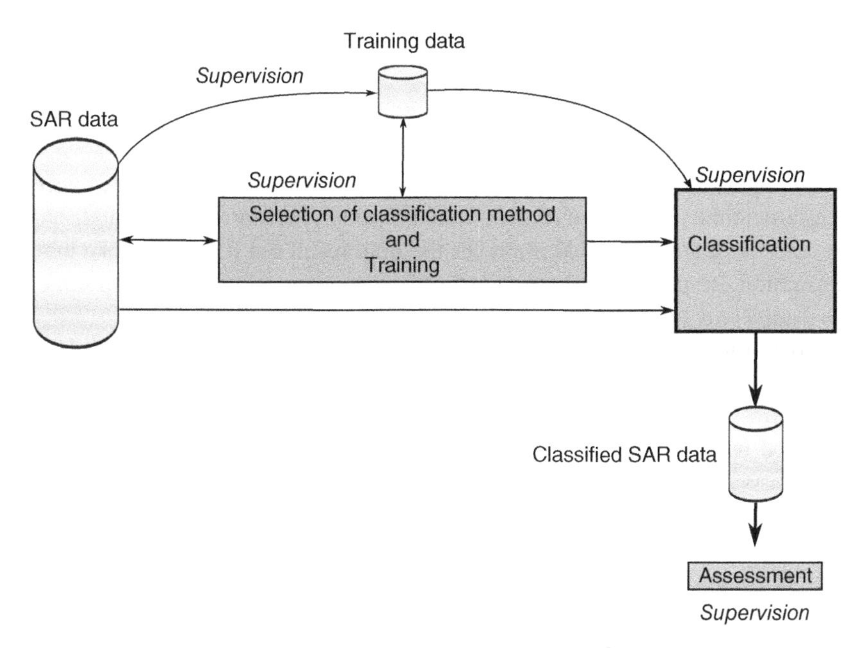

Figure 5.35 Supervised SAR image classification process. Supervision by the expert is done in many stages of the process.

of SAR data (especially single look data). Most of the methods reach a tradeoff by slightly filtering the original data (a median filter with small convolutional size, for example, of radius 3 or better by using a despeckling filter). In the results shown in this section, the result for filtered data and for not filtered data are shown.

5.6.1 The Nearest Neighbor Classifier

The first approach to classify an image is a *naive* one: the nearest neighbor classifier. As its name suggests, it aims to classifying pixels by minimizing the Euclidean distance among the feature vectors. It works like,

- the user indicates the number of classes,
- features are defined (for instance, grayscale values or texture descriptors for each pixel within the SAR image), and at least a sample is assigned to each class (training phase),
- let the feature vector of the jth sample within the class i^{th} be $\mathbf{x}_{i,j}$,
- to classify a new unknown vector $\mathbf{x}$ (a SAR pixel) and by using the Euclidean distance, its nearest neighbor is found and x is assigned to the class that it belongs to.

Furthermore, for some k and ℓ indices of the data,

$$|\mathbf{x}_{k,\ell} - \mathbf{x}| < |\mathbf{x}_{i,j} - \mathbf{x}|, \forall i \text{ and } \forall j, \tag{5.21}$$

then, $\mathbf{x}$ is assigned to class k.

This *naive* approach is not recommended because

- for overlapped classes, the assignment to the correct class is not guaranteed, and resembles almost randomly (it depends on the proximity of the the initial selected sample for each class),

- if not an intelligent distance calculation and distances stored is done, the method is computationally expensive.

Additionally, noise (speckle) largely degrades the performance of this simple method that, if properly programmed works well for particular problems.

Nearest neighbor algorithm is implemented in R (package 'neighbr', https://cran.r-project.org/web/packages/neighbr/neighbr.pdf) and in Matlab through the *knnsearch* method.

As pointed out above, filtering notably improves the final result but it may provoke loss of valuable information. Results shown below are for filtered and non-filtered data. In some particular cases results only for non-filtered or for filtered data are discussed.

Figure 5.37 shows the result from the nearest neighbor classifier for simulated data (ENL = 1), which is a *worst-case* example. The training set selected for each class is also included in this figure. The result is not particularly good, but, this is a worst-case and the classification method used is a basic one. However, even for the non-filtered case shown, some classification was achieved and it is visually noticed. For the filtered data, classification has improved notably. Of course, the initial solution (the result from the despeckling filter) has a strong impact on the final classified image. Hence, if instead of applying the Lee filter, which is an excellent SAR despeckling filter, other more updated filter was used, the classified result, even for this simple classifier, would enhance accordingly.

A similar result for ENL = 3 is depicted in Figure 5.38, and as it can be seen, the classification result has improved with respect to the previous result due to the speckle content is less.

Results shown were obtained using Matlab with the script available in file `Code/Matlab/Chapter6/Nearest_neighbor.m` from www.wiley.com/go/frery/sarimageanalysis.

To reduce computational complexity due to the estimation of distances from the unknown data to be classified to all samples within training sets, the nearest-centroid classification method is applied. As its name suggests, only the distance to the centroid (center of mass) of the training sets is used to perform the classification. In doing that, a great improvement in storing data and computing distances is obtained. This method is quite simple, easy to implement and it works well for well-balanced clusters (data similarly distributed) that do not overlap excessively.

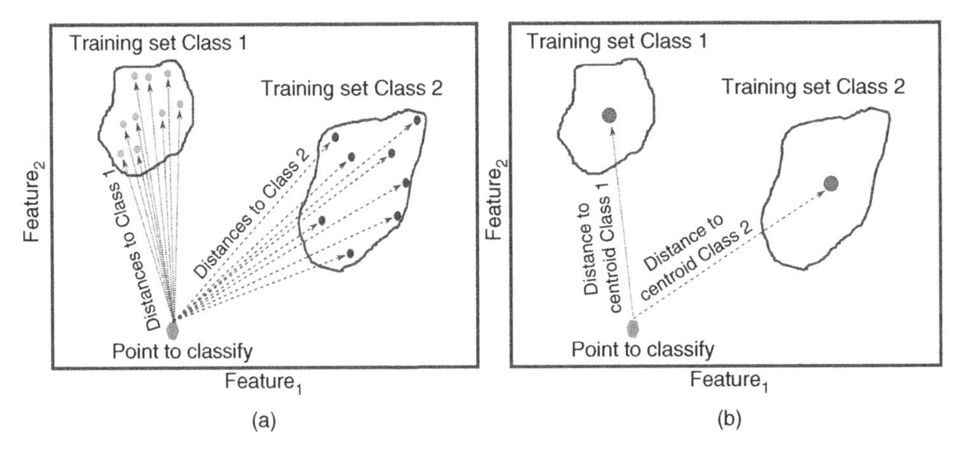

Figure 5.36 Nearest neighbor classification (a) vs. nearest centroid classification (b). For the latter, only two distances among features are calculated.

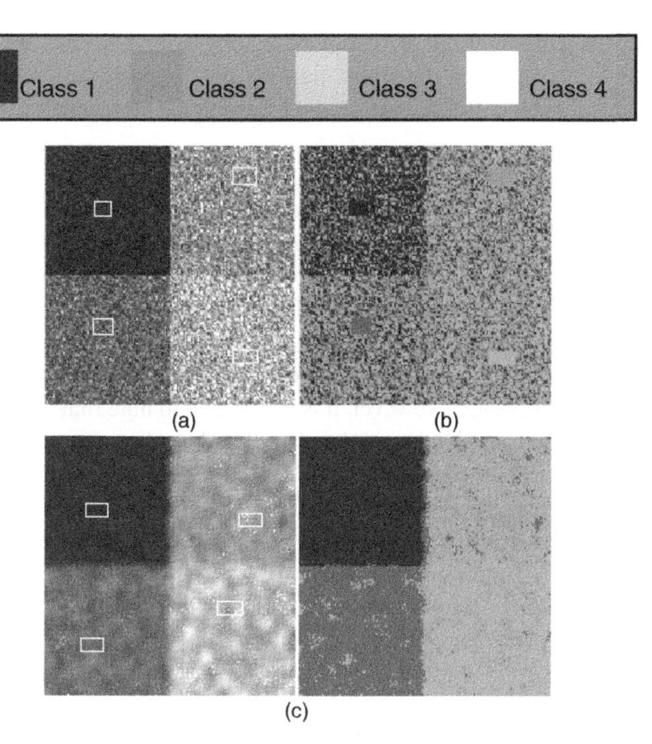

Figure 5.37 Nearest neighbor classification for simulated data (ENL = 1). (a) the simulated data and the classification result (b). In (c), the classification method is applied to the filtered data (by using the Lee filter). The training set for each class is also shown (white rectangle within each class).

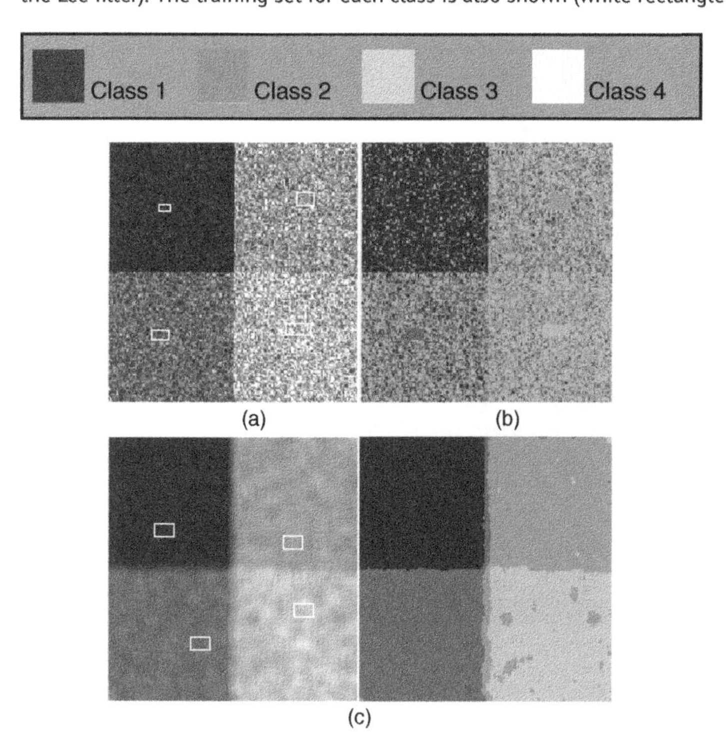

Figure 5.38 Nearest neighbor classification for simulated data (ENL = 3). (a) the simulated data and the classification result (b). In (c), the classification method is applied to the filtered data (by using the Lee filter). The training set for each class is also shown (white rectangle within each class).

Figure 5.36 shows how the computation of distances among points in the feature space gets strongly reduced when replacing the whole training sets for each class with its unique representative features. Each centroid is calculated by averaging all features within each class.

Figures 5.39 and 5.40 show the results for two simulated SAR data (ENL = 1 and ENL = 3). As it can be seen, the results are not much different than the ones obtained with the nearest neighbor method. It is easy to identify the solution for the nearest neighbor method just looking the training areas within the classified image, resembling well-classified, as expected (perfect due to they were selected in the training phase, so they were already classified by the user).

Figure 5.41 shows the result the San Francisco Bay data (not filtered). Four classes (ocean, forest, urban, and grass) have been selected. However, it is interesting to note that the quality of the result depends strongly on the selection of the classes and on the selection of the training set. For this SAR data, the training set has been carefully selected due to a poor selection provides a too bad result (not shown). As it can be seen, some areas of the ocean have been wrongly classified as grass due to the backscatter values are similar in some pixels for the two areas (ocean and grass). Therefore, the selection of classes (and features) plays a key role in the classification process. However, the result from this simple classification method is acceptable for all classes. The nearest centroid method is also very fast and easy to implement.

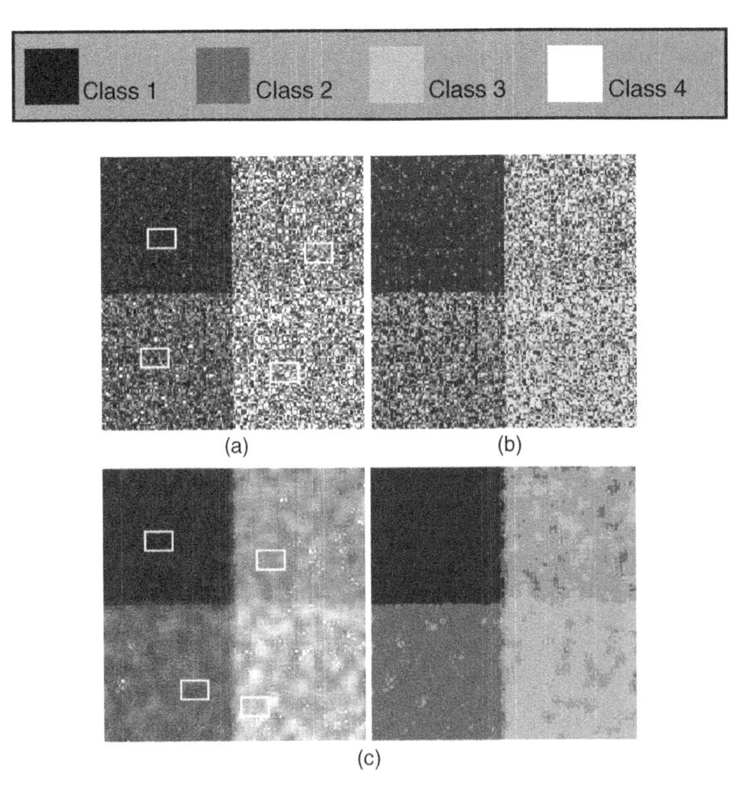

Figure 5.39 Nearest centroid classification for simulated data (ENL = 1). (a) the simulated data and the classification result (b). In (c), the classification method is applied to the filtered data (by using the Lee filter). The training set for each class is also shown (white rectangle within each class).

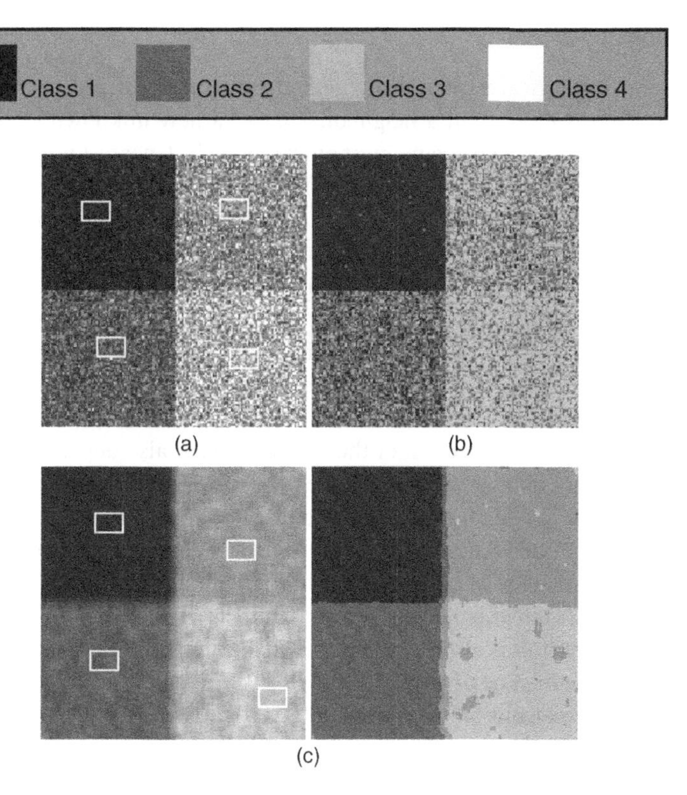

Figure 5.40 Nearest centroid classification for simulated data (ENL = 3). (a) the simulated data and the classification result (b). In (c), the classification method is applied to the filtered data (by using the Lee filter). The training set for each class is also shown (white rectangle within each class).

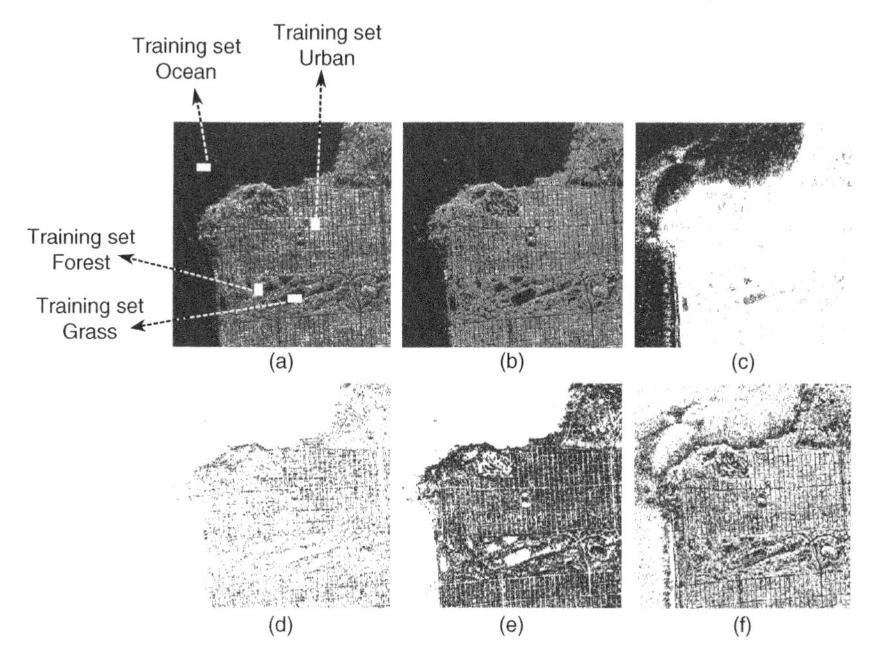

Figure 5.41 Nearest centroid classification for San Francisco Bay (not filtered data). (a) the original image, the classified result (b) and the binary masks (in black color) for the classified ocean (c), the classified urban area (e), the classified forest area (d) and, the classified grass area (f). The training set for each class is also shown (white solid rectangle within each class).

5.6.2 The K-nn Method

The natural improvement of the nearest neighbor method consists of instead of look at the nearest neighbor, to look to more than a neighbor. That is, a new unknown element (pixel), **x**, will be assigned to the class that occurs most often among its k nearest neighbors[5]. The k-nearest neighbors method (k-nn), is a well-established nonparametric classification method in pattern recognition and machine learning and, the nearest neighbor discussed above is the k-nearest neighbors method for $k = 1$. K-nn is also available in the R and Matlab packages mentioned above. The basic idea behind the K-nn method is to get local information for already classified data. Figure 5.42 illustrates this with a two-class simple example, where an unknown data must be classified as belonging to Class 1 or to Class 2. If $k = 1$ (nearest neighbor), the new data will be assigned to Class 2, if $k = 3$, the new data will be assigned to Class 1 and for $k = 5$, the new data will be assigned to Class 1.

Therefore, the result depends on the election of the k value, which also depends on the distribution of the data. There are many recommendations to assist in the election on k and in a practical way, the user will run several experiments to find a suitable k value.

The explanation of the K-nn method requires also to mention that other distances than the Euclidean can be used (for instance the Mahalonabis distance), and that data is usually normalized to avoid biased for classes too unbalanced (different number of elements). Weighting the distances according to a criterion is also convenient to make K-nn more robust against variation in distances of the k-nearest neighbors.

Figures 5.43 and 5.44 show the results for two simulated SAR data (ENL = 1 and ENL = 3). As it can be seen, the result is quite similar than for the $k = 1$ case (nearest neighbor method) and this clearly shows that not always increasing the number of neighbors (pixel neighbors) a better result is obtained.

Figure 5.45 shows the result for the San Francisco Bay data (not-filtered data). Four classes (ocean, forest, urban, and grass) have been selected. An appreciable improvement is visible when compared to the previous classification by the nearest centroid method. If the method were applied to the filtered data, the final classification result will largely improve as it has been shown for the simulated data discussed above.

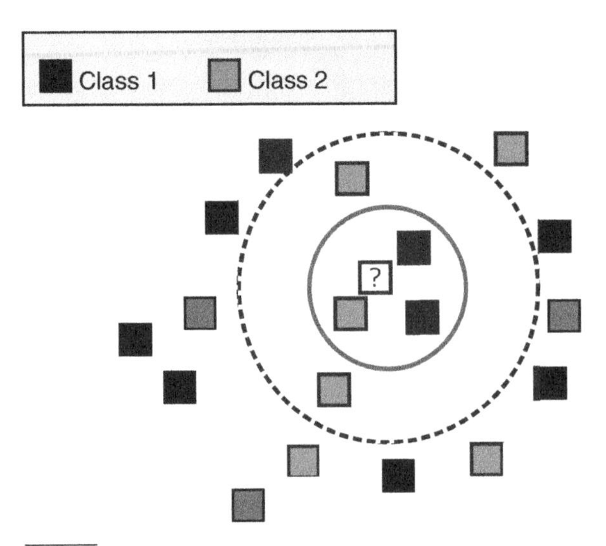

Figure 5.42 k nn classification for $k = 3$ and $k = 5$. The representation is in the feature space and the Euclidean distance is used.

5 In the literature, k is used to design the neighbors.

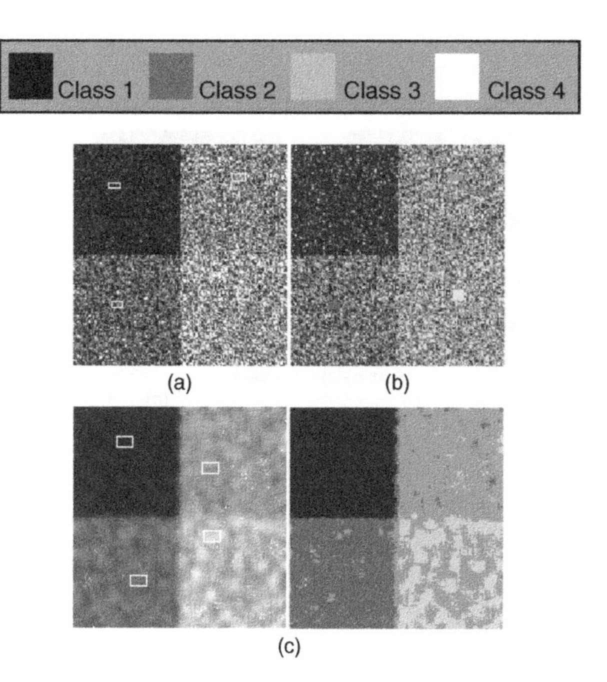

Figure 5.43 K-nn classification for simulated data (ENL = 1), with k = 5. (a) the simulated data and the classification result (b). In (c), the classification method is applied to the filtered data (by using the Lee filter). The training set for each class is also shown (white rectangle within each class).

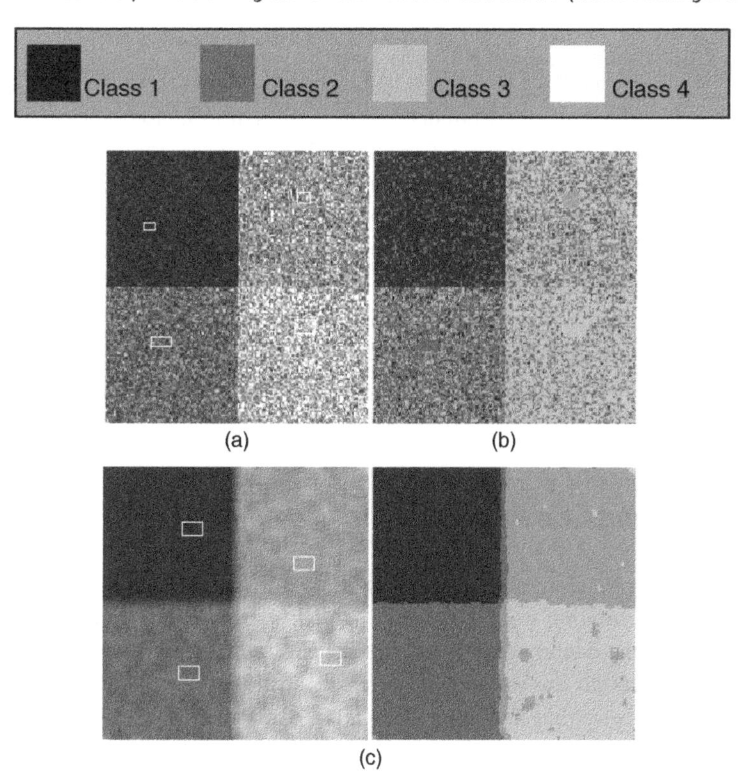

Figure 5.44 K-nn classification classification for simulated data (ENL = 3), with k = 5. (a) the simulated data and the classification result (b). In (c), the classification method is applied to the filtered data (by using the Lee filter). The training set for each class is also shown (white rectangle within each class).

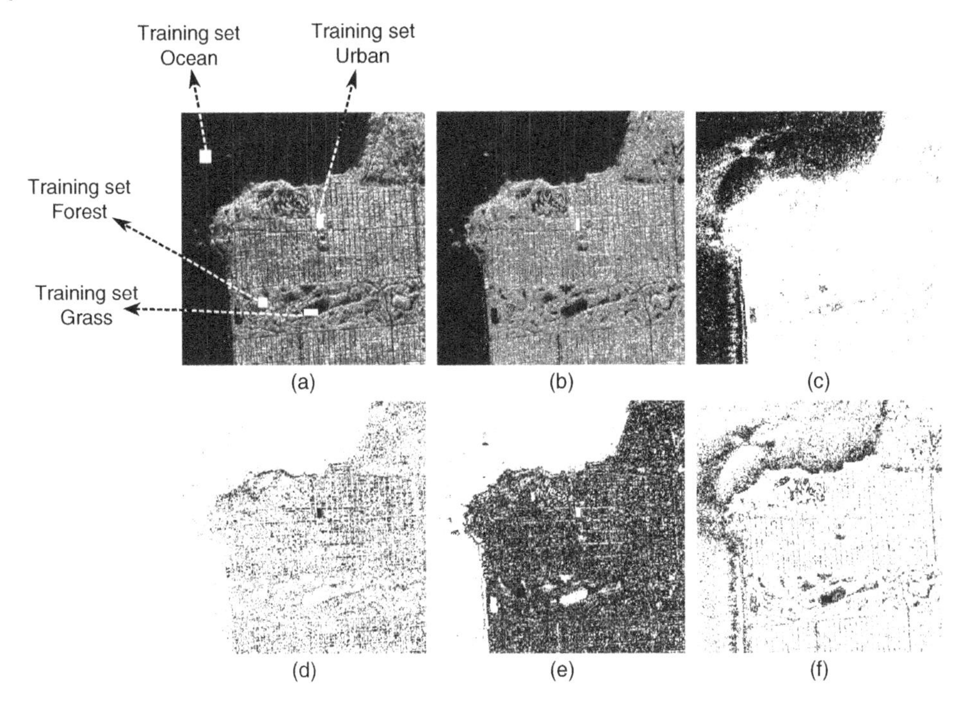

Figure 5.45 K-nn classification for San Francisco Bay (not filtered data), with k = 5. (a) the original image, the classified result (b) and the binary masks (in black color) for the classified ocean (c), the classified urban area (d), the classified forest area (e) and, the classified grass area (f). The training set for each class is also shown (white solid rectangle within each class).

Additionally, for all the examples shown (simulated pattern and San Francisco Bay), the single-pixel value has been selected as the feature. Obviously, this feature is extremely simple and, better results can be obtained by selecting other more descriptive features (Haralick's descriptors or others). It is also important to note that the computational times for the methods discussed herein are not affected significantly by using more complex descriptors due to the only impact on the computational times lies on the calculus of the distances (Euclidean distances for the cases analyzed).

Results shown were obtained using Matlab with the script available in file `Code/Matlab/Chapter6/Nearest_neighbor.m` from www.wiley.com/go/frery/sarimageanalysis, and the number of neighbors was larger than 1.

5.7 Maximum Likelihood Classifier

Another simple and efficient approach to classify data is the Maximum Likelihood method (ML), which aims to extract information from data from a statistical point of view. In doing so, this method considers both the center of clusters (through their mean values) and their shape, size, and orientation (through the variance or covariance values). The main idea of the ML is to predict the class label $\mathbf{y}$ that maximizes the likelihood of the observed data $\mathbf{x}$. This is done by estimating a statistical distance from the mean values and covariance matrix of the clusters. Therefore, a first assumption is that the statistics of the clusters follow a Gaussian (normal) distribution, which, in SAR data can be accepted due to the inherent random distribution of scatterers within the resolution cells. Once the probabilistic distribution functions, *pdf*, for each class is known, the pixel to be classified is assigned to the class (cluster) to which it has the highest probability. ML, jointly with the previous method above explained, is a well-known classification method that is also simple to understand and to implement, providing acceptable results at least to have an initial classification for others more complex methods to start working on. It is available in most mathematical packages (see https://search.r-project.org/CRAN/refmans/rasclass/html/rasclassMlc.html for R and the methods *mle* and *mvnpdf* for Matlab).

ML classifier operates in this way,

Learning phase:

- the user defines the number of classes and provides some training data for each class,
- for each class **y**, the parameters of the normal *pdf* are estimated from the training data (mean values and the covariance values).

Assignation phase:

- each pdf is evaluated for each x to be classified,
- each **x** is assigned to the class y that maximizes the evaluated *pdf* (likelihood of the observed data).

Figure 5.46 shows a 1-dimensional example for a two-classes problem (Class 1 and Class 2). As it can be seen, from the estimation phase, the pdf of each class is available (pdf_{C1} and pdf_{C2}), which correspond to data distributed with estimated $\mu = 1.5$, $\sigma = 0.5$ and $\mu = -1$, $\sigma = 1$, respectively. In the assignation phase, the data to be classified ($\mathbf{x} = \mathbf{0}$) have $pdf_{C1}(\mathbf{0}) = 0.0088$, and $pdf_{C2}(\mathbf{0}) = 0.2419$, so, as $pdf_{C1}(\mathbf{0}) > pdf_{C2}(\mathbf{0})$, $\mathbf{x} = \mathbf{0}$ is assigned to Class 2.

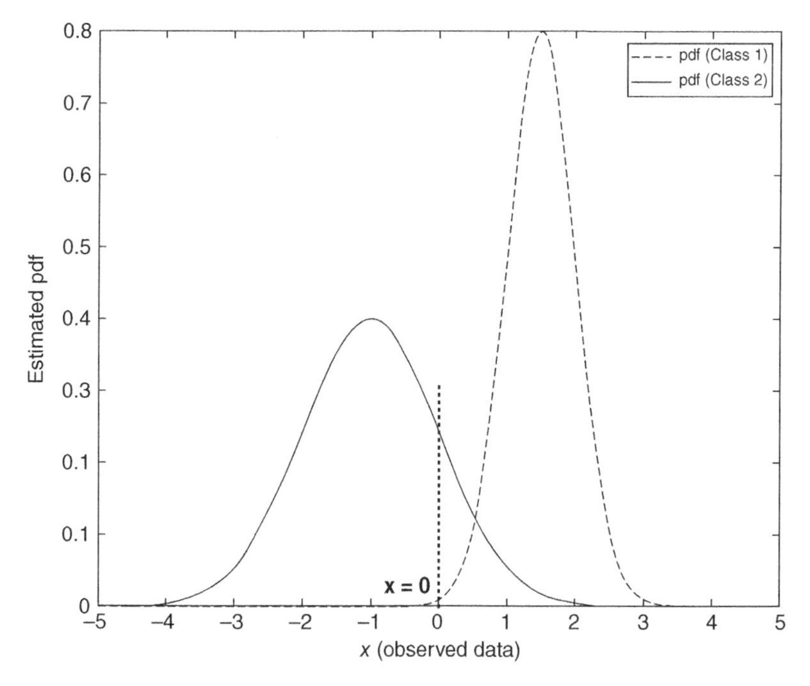

Figure 5.46 Example of ML two-classes problem classification using a single descriptive feature.

For actual SAR data, and in general for most practical problems, a single feature (vector of dimension 1) is not enough to get a reasonable result, hence, the 1-dimension Gaussian *pdf* must be replaced by the multivariate Gaussian *pdf*,

$$f(\mathbf{x}) = \frac{1}{\sqrt{(2\pi)^d \det(\Sigma)}} \cdot e^{-\frac{1}{2}(\mathbf{x}-\mu)^T \Sigma^{-1}(\mathbf{x}-\mu)}, \tag{5.22}$$

where $\mathbf{x}$ is a column vector (data from one observation), d is the dimension of $\mathbf{x}$ ($\mathbf{x}$ is a $d \times 1$ vector), μ is the estimated mean value of $\mathbf{x}$ (it is also a $d \times 1$ vector) and Σ is the estimated covariance matrix of $\mathbf{x}$ (a $d \times d$ matrix). The covariance matrix, Σ, should be positive definite (to assure that its determinant, $\det(\Sigma) = |\Sigma| \neq 0$), so, Σ is symmetric and all eingevalues should be positive.

Besides, it is common to include a threshold value (threshold distance) by defining a maximum probability value (equivalent to confidence intervals). This must be done carefully to avoid inter-class biases.

The case of using two features in a two-classes problem is illustrated in Figure 5.47. Each feature, for each class, C_1 and C_2, has a mean value μ_1, μ_2, and a standard deviation σ_1, σ_2. The spatial distribution of data into the feature space is taken into account during the classification as it can be seen from the spatial location of the elliptic areas (ellipsoids in more dimensions). The covariance matrix Σ accounts for the spatial dispersion of features within each class.

Below, some examples for the ML classification of actual SAR data are shown.

Figures 5.48 and 5.49 show the results for two simulated SAR data (ENL = 1 and ENL = 3). As it can be seen, the results are not much different than the ones obtained with the previous classification methods.

Figure 5.50 shows the result for the San Francisco Bay data (not-filtered data). Once again, four classes (ocean, forest, urban, and grass) have been selected.

For all these three examples shown, two features have been selected, the pixel gray level and the coefficient of variation. The latter has been calculated pixelwise and it is given by,

$$CV = \frac{\sigma_{local}}{\mu_{local}}, \tag{5.23}$$

where σ_{local} is the standard deviation locally estimated at each pixel (a neighborhood of 3×3 has been applied) and, μ_{local} is the mean value estimated at each pixel (a neighborhood of

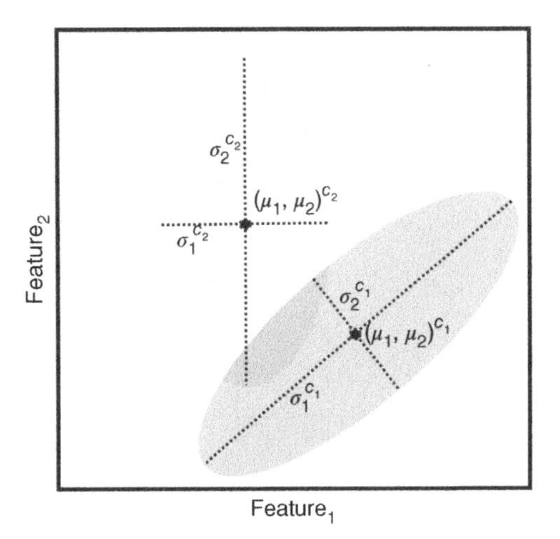

Figure 5.47 Example of ML two-classes problem classification using two descriptive features.

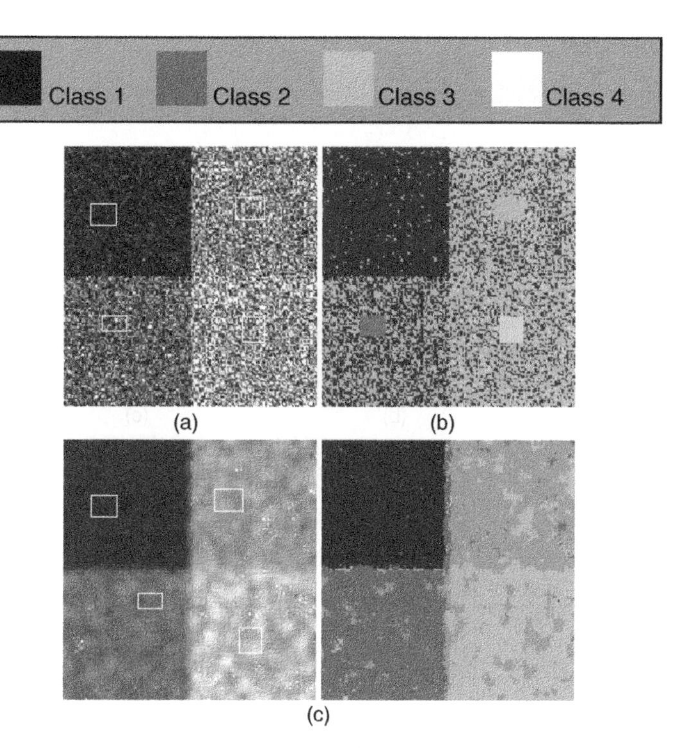

Figure 5.48 Maximum likelihood classification for simulated data (ENL = 1). (a) the simulated data and the classification result (b). In (c), the classification method is applied to the filtered data (by using the Lee filter). The training set for each class is also shown (white rectangle within each class).

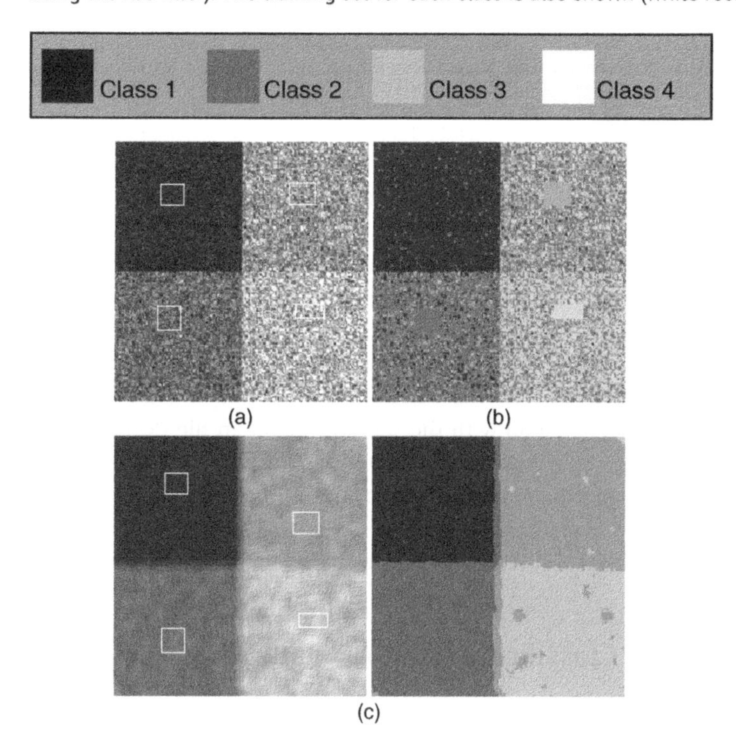

Figure 5.49 Maximum likelihood classification for simulated data (ENL = 3). (a) the simulated data and the classification result (b). In (c), the classification method is applied to the filtered data (by using the Lee filter). The training set for each class is also shown (white rectangle within each class).

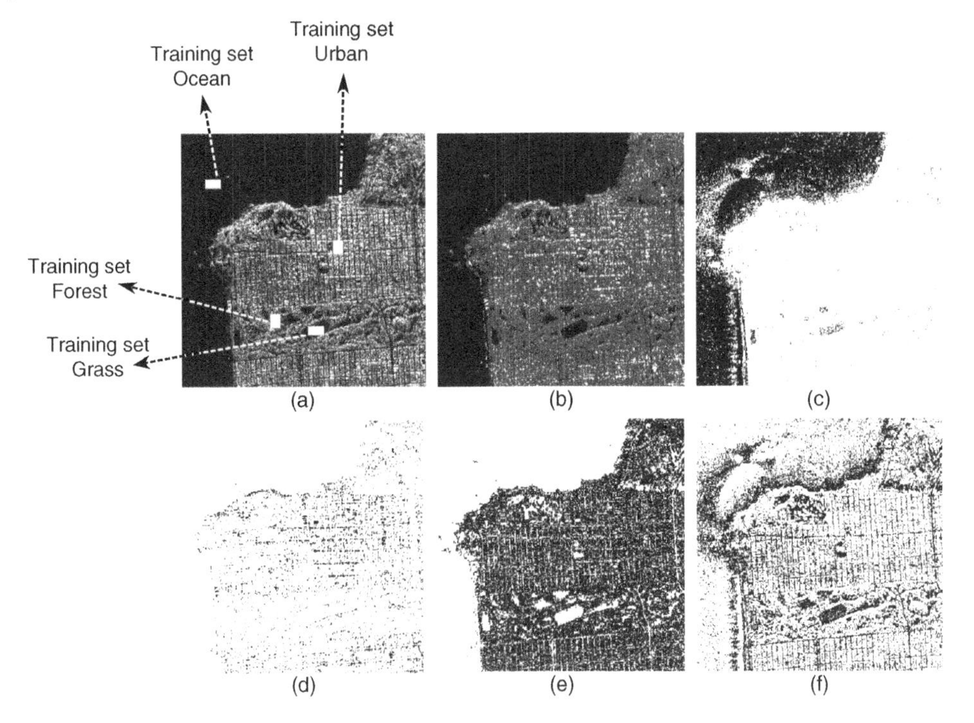

Figure 5.50 Maximum likelihood classification for San Francisco Bay (not filtered data). (a) the original image, the classified result (b) and the binary masks (in black color) for the classified ocean (c), the classified urban area d), the classified forest area (e) and, the classified grass area (f). The training set for each class is also shown (white solid rectangle within each class).

3×3 has been applied). Therefore, the feature vector $\mathbf{v} = \left(a_{i,j}, CV_{i,j}\right)$, where i and j account for the pixel coordinates in the image space.

The multivariate Gaussian *pdf* reduces to a bivariate distribution, with $\mu = \left(\mu_1, \mu_2\right)$, where μ_1 and μ_2 are the estimated mean values for each training set for the two selected features, $a_{i,j}$, CV. The estimated covariance matrix, Σ is estimated and it is a 2×2 matrix. Each observation data $\mathbf{x}$ is a 2×1 vector.

It can be seen that by using more features the classification notably improves (see the classified image, top row, center image and compare it with the corresponding ones for the previous methods). However, the classification of the grass class is still poor due to the reason given above: not many available samples for training the method and their similarity to the backscatter values for the ocean area.

Results shown were obtained using Matlab with the script available in file `Code/Matlab/Chapter6/ML_classifier.m` from www.wiley.com/go/frery/sarimageanalysis.

ML method provides excellent results for images with little noise, which is not the case with SAR data. However, as it has been shown, the result for the actual SAR data (not-filtered) is acceptable. Besides, the computational cost is almost neglectable even for large SAR data.

5.8 Unsupervised Image Classification of SAR Data: The K-means Classifier

Supervised classification implies the user provides enough information of the area to classify. If that information is not enough, supervised methods can not be applied or if so, the

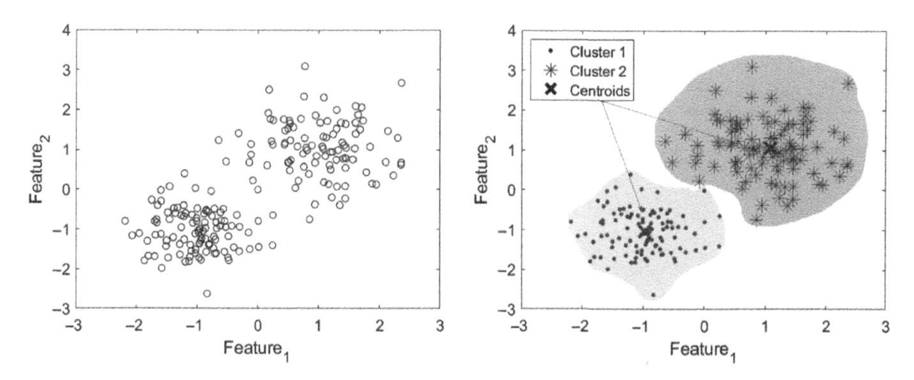

Figure 5.51 Example of K-means two-classes problem classification using two descriptive features.

quality of the final result would not satisfy the user's necessities. The option is to apply an unsupervised classification method. Besides, occasionally, data will distribute naturally within well-separated clusters, inviting to use automatic (unsupervised) methods. The basic idea behind an unsupervised classification method is to iteratively merge existing clusters. Although no-user intervention is formally required, in practical situations, the user has a minimum knowledge of the data and provides the desirable number of classes. From that, centroids of each cluster are calculated as an initial step of the process, and points are first assigned to a cluster according to a similarity criterion (minimum Euclidean distance in the feature space) into clusters (*labeling* of all points) and, at each iterative step of the classification process, two clusters that contain two points that are closer together than any other two points in distinct clusters are merged. Iterations continue until, either the final number of clusters is the desired one (cluster centers do not change) or, when a new point to be included in a cluster does not satisfy the minimum distance criterion (its distance is more than some threshold). Some refinements are needed to avoid excessive partition of the areas. The usual one is to remove -at any iteration- clusters with less than a specified number of points or to merge clusters that are less separated than a pre-defined threshold between cluster centroids. It is interesting to remark that the final classification obtained is indeed described by the data distribution on the feature space.

Figure 5.51 shows an example of classification by using the well-known K-means method (do not confuse with the K-nn classification method discussed in Section 5.6.2). As seen, for this example with two well-separated data, the final two clusters group most of their nearest points into two clusters (cluster 1 and cluster 2).

The K-means is considered one of the most popular unsupervised clustering algorithms. It is easy to understand and code, and it provides acceptable results at a low computational cost. From a mathematical perspective, the K-means method aims to find an optimal solution following an expectation-maximization approach. In the expectation phase, points are assigned to their closest cluster. In the maximization phase, the centroid of each cluster is updated.

As it happens with the method that uses a similarity criterion based on distances, it is recommended to normalize the data to have a mean of zero and a standard deviation of one to suit different units of measurements.

It is also recommended to run the K-means method several times to avoid the algorithm to stuck at a local minimum. This is due to the random initialization of centroids at the start of the method: different initializations may lead to different final results. Therefore, the user must run several times the algorithm and select the result showing the lower sum of squared distances. This is not, in general, a problem due to computational cost is low.

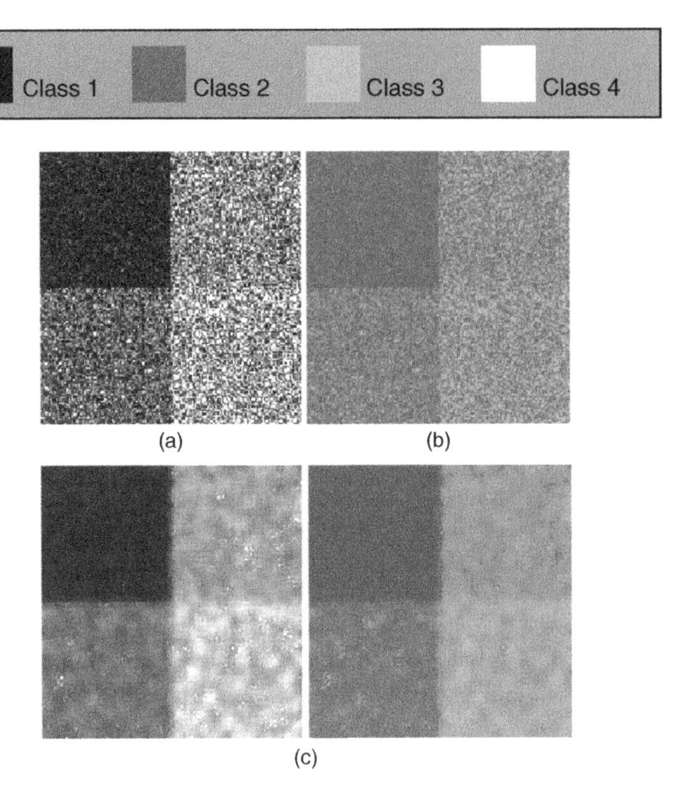

Figure 5.52 K-means classification for simulated data (ENL = 1), with four initial clusters. (a) the simulated data and the classification result (b). In (c), the classification method is applied to the filtered data (by using the Lee filter).

K-means method is included in most mathematical packages. It is available in the statistical R package through the *kmeans* method (see https://www.rdocumentation.org/packages/stats/versions/3.6.2/topics/kmeans) and in the *kmeans* function in Matlab).

Figures 5.52 and 5.53 shows the classification result by using the K-means method for two SAR simulated data. Once again, results for noisy data are poor and notably improve for the filtered data (by the Lee filter). In this case, results for the filtered data are also much better than the ones obtained by the previous methods.

Results shown were obtained using Matlab with the script available in file `Code/Matlab/Chapter6/Examples_kmeans.m` from www.wiley.com/go/frery/sarimageanalysis.

To conclude this section devoted to simple classification methods, it can be said,

- supervised methods require user intervention to defining the number of classes and to provide a training set for each class (hence, descriptive features must be also provided by the user),
- unsupervised methods do not require user intervention beyond providing the number of final clusters (or classes),
- unsupervised methods are not indicated for classes with a significant overlapped,
- the simple methods discussed in this section were indeed designed to classifying natural images or images with little noise. Yet they might provide acceptable results for SAR images in some cases,

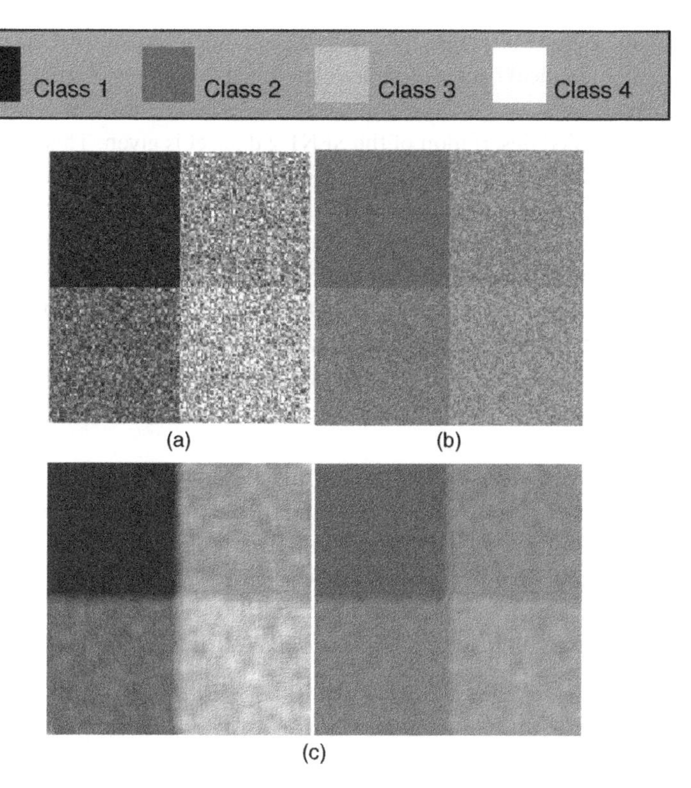

Figure 5.53 K-means classification for simulated data (ENL = 3), with four initial clusters. (a) the simulated data and the classification result (b). In (c), the classification method is applied to the filtered data (by using the Lee filter).

- in general, not a method is superior to other, and sometimes, the simpler one can better suit the user necessities.

These simple methods herein discussed can be seen as a first approach to launch more sophisticated methods beyond the scope of this book.

5.9 Assessment of Classification Results

Assessment of a classification technique is done simply by *counting* the number of classified data (pixels) that should belong to each class. Obviously, the *true* total data belonging to each class must be 'a priori' known. Sometimes it is not required to count all pixels but sampling some areas. For example, for the case of the simulated data used in previous examples (the 150×150 four classes pattern), each class has a total of $75 \times 75 = 5625$ points. If the classification result for the first class gives a total of 5625 classified points, the error in this class has been zero. From the obtained number of classified pixels, many metrics can be built (see below). But, before introducing the metrics, an important issue related to SAR data is remarked: in most cases, there is no available a ground truth to compare the classified pixels with. In this case, the usual approach consists of making an enough-valid ground truth. This is done just by using an optical image corresponding to the SAR data to classify and manually marking the classes. To do this, both images must be co-registered, that is, an affine transformation between the two images may be obtained. Such transformation may be even nonlinear. Most mathematical packages include methods to co-register images, so, this should not be a problem.

In Schmitt et al. (2018) a detailed description of the SEN1-2 dataset is given. This dataset (shared under the open-access license CC-BY), available for download in https://mediatum.ub.tum.de/1436631 contains a relatively large amount of remote sensing data from SAR data (Sentinel-1) and optical data (Sentinel-2) co-registered and ready for its use. The intended application is, as it is mentioned in the article, to training deep learning methods (for instance to artificially generate RGB images from SAR data by fusing radar and optical images). For testing classification methods, its use is strongly recommended to have a ground truth to compare classification results with. Some pair of images from this dataset are shown in Figures 5.54 and in 5.55. The optical images are RGB images, shown in this book as grayscale images.

To evaluate a classification method with a ground truth available, the recommended approach is to make an image with a mask for each class and then evaluate the metrics. Two examples of such masks (from Uhlmann and Kiranyaz (2014)) for the Flevoland and the San Francisco Bay images are shown in Figures 5.56 and 5.57 respectively.

The most used approach for assessing a classification result is to organize the counting of classified pixels per class in a table known as *confusion matrix*. A two-class classification problem is depicted in Figure 5.58 and its confusion matrix after classification by a method is shown in Table 5.6.

The main diagonal indicates the *true positives*, TP, that is, observations well classified (there are 90 observations out of a total of 100 well classified for the Class 1 and 75 out of a total of 100 well classified for the Class 2). It is expected for any classification method to provide large true positives, which is easily checked out just by looking to the values in the main diagonal of the confusion matrix (usually in bold fonts due to its relevance).

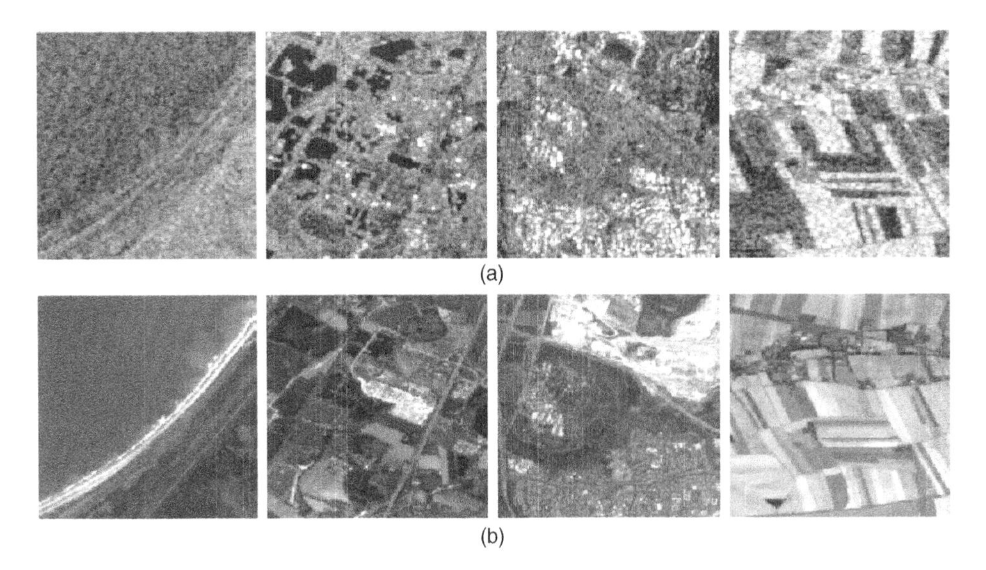

Figure 5.54 Some exemplary patch-pairs from the SEN1-2 dataset (from Schmitt et al. (2018)). (a) shows the Sentinel SAR images and, (b) the Sentinel-2 RGB images. The original Sentinel-2 images are RGB images, shown in this book as grayscale images.

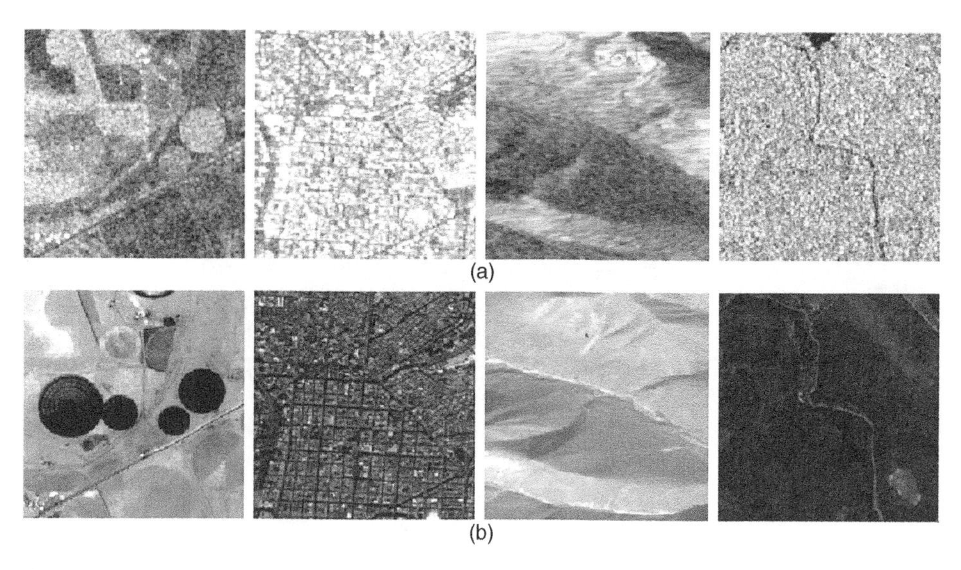

Figure 5.55 Some exemplary patch-pairs from the SEN1-2 dataset (from Schmitt et al. (2018)). (a) shows the Sentinel SAR images and, (b) the Sentinel-2 RGB images. The original Sentinel-2 images are RGB images, shown in this book as grayscale images.

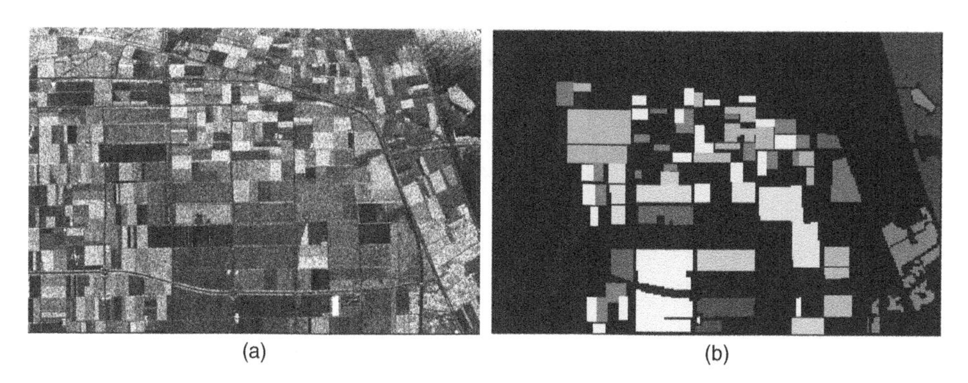

Figure 5.56 (a) Flevoland HH AirSAR image and (b) ground truth manually marked from optical image. The description of each mask (original in color) is detailed in Uhlmann and Kiranyaz (2014).

Another important information shown in the confusion table is given by the false positives, FP, (false alarm or what is known in statistical hypothesis test as *Type I error* or overestimation), the false negatives, FN, (miss or what is known in statistical hypothesis test as *Type II error* or underestimation), the true negatives, TN (correct rejection), and the true positives, TP (correct hit). This information is contained in the off-diagonal cells and indicates wrongly classified observations. These concepts (FP, FN, TN, TP) measure the sensitivity and the specificity from the classification obtained (see Table 5.7).

From the information contained in the confusion matrix, *Accuracy* is calculated as,

$$Accuracy = \frac{TP + TN}{TP + TN + FP + FN}, \tag{5.24}$$

and it somehow measures the proportion of correct guesses in the classification obtained.

The last metric explained herein is the known as the kappa measure, κ, which is commonly added to the confusion matrix and it resumes the global classification result. κ is designed

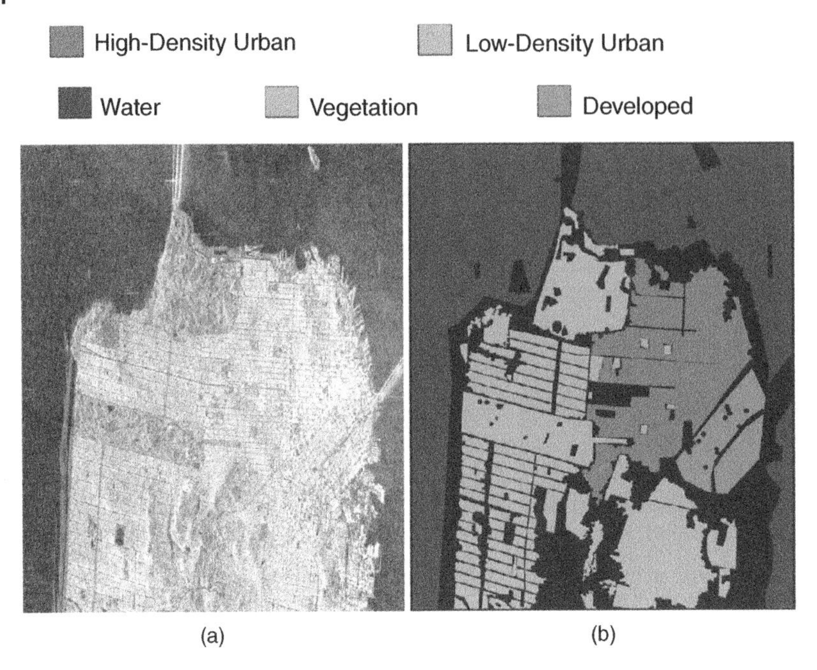

Figure 5.57 (a) HH band San Francico Bay image and (b) ground truth manually marked from optical image. The description of each mask (original in color) is detailed in Uhlmann and Kiranyaz (2014).

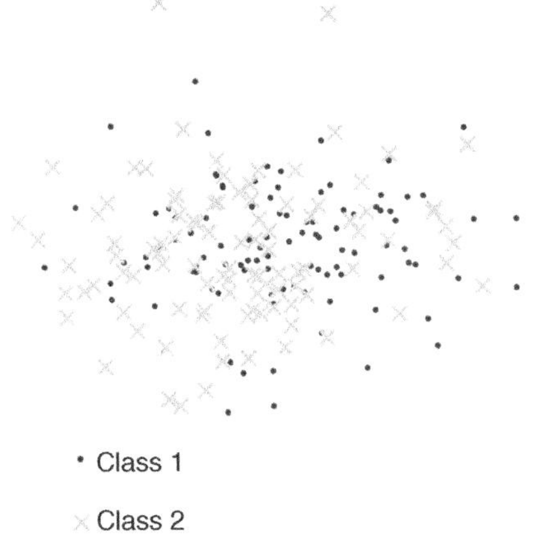

Figure 5.58 Two-classes classification problem (there are 100 observations for each class).

with the purpose to assure that the performances of the classification method is not due to random classification. It must be noted that in any classification method, some observations are well classified simply due to random classification, what might happen more in data corrupted with random noise. κ is a single-value metric and it is calculated as follows,

$$\kappa = \frac{Accuracy - Accuracy^{random}}{1 - Accuracy^{random}}, \tag{5.25}$$

where $Accuracy^{random}$ aims to measure the potential random classification and it is calculated as,

$$Accuracy^{random} = p_1 p_2 + (1 - p_1)(1 - p_2), \tag{5.26}$$

Table 5.6 Confusion matrix for a two-class classification problem.

True Class \ Predicted Class	Class 1	Class 2
Class 1	**90 (90%)**	10 (10%)
Class 2	25 (%25)	**75 (75%)**

Table 5.7 Sensitivity and specificity for the two classes classification example.

True Class \ Predicted Class	Class 1	Non Class 1
Class 1	**90 TP**	10 FN
Non Class 1	25 FP	**75 TN**

where,

$$p_1 = \frac{TP + FN}{TP + FP + TN + FN}, \tag{5.27}$$

and,

$$p_2 = \frac{TP + FP}{TP + FP + TN + FN}. \tag{5.28}$$

It is clear that $\kappa \leq 1$ and it is $\kappa = 1$ when the Accuracy is 1.

The two-classes confusion matrix detailed above can be applied to analyze a multi-class classification problem.

Tables 5.8 and 5.9 show the confusion matrix for the classification with the K-nn method (5 neighbors), of the simulated $ENL = 1$ SAR data (the *pattern* used in previous examples), non

Table 5.8 Confusion matrix for the simulated ENL = 1 data (pattern) for the K-nn (5 neighbors) applied to the non filtered data.

True Class \ Predicted Class	Class 1	Class 2	Class 3	Class 4
Class 1	**3837 (68.21 %)**	1477 (26.25 %)	874 (15.53 %)	656 (11.66 %)
Class 2	153 (2.72 %)	**412 (7.32 %)**	324 (5.76 %)	225 (4 %)
Class 3	997 (17.72 %)	2142 (38.08 %)	**2377 (42.25 %)**	2084 (37.04 %)
Class 4	638 (11.34 %)	1594 (28.37 %)	2050 (36.44 %)	**2660 (47.28 %)**

NOTE: A total of 5625 points are considered per class (75 × 75).

Table 5.9 Confusion matrix for the simulated ENL = 1 data (pattern) for the K-nn (5 neighbors) applied to the filtered data (Lee filter).

True Class \ Predicted Class	Class 1	Class 2	Class 3	Class 4
Class 1	**5479 (97.40 %)**	2 (0.035 %)	1 (0.017 %)	0 (0 %)
Class 2	146 (2.59 %)	**5195 (92.35 %)**	254 (4.51 %)	9 (0.16 %)
Class 3	0(0 %)	395 (7.02 %)	**3910 (69.51 %)**	1254 (22.29 %)
Class 4	0(0 %)	33 (0.58 %)	1460 (25.95 %)	**4362 (77.54 %)**

NOTE: A total of 5625 points are considered per class (75 × 75).

filtered and, filtered with the Lee filter respectively. As was expected from the visual analysis performed above, filtering significantly improves the classification results. Moreover, the confusion matrix allows to inspect what classes have been best classified, that is, it provides clues to improve the classification performed.

Exercises

1. Obtain X_{MAP} under the $\mathcal{K}$ model for the return.

2. Implement the MAP filters derived in Section 5.2.1 and in Exercise 1. Apply them to simulated data, and compare the results. Try a variety of window sizes, and of estimators.

3. The codes provided for all the classification methods were coded in Matlab. From the experience in `R` achieved from this book, and following these Matlab codes is easy to code them into `R`. This is a recommended exercise that enforces all the acquired knowledge in `R` programming.

4. Select two filters and two metrics and run a Monte Carlo analysis to compare their performances.

5. Run a Monte Carlo analysis, select an actual SAR image and a metric and tune the decay parameter, h for the NLM filter.

6. Run a Monte Carlo analysis, select an actual SAR image and a metric and tune the η parameter, h for the NLM-Statistical filter.

7. Run a Monte Carlo analysis, select an actual SAR image and a metric and decide what stochastic distance performs the best for the NLM-Statistical filter.

8. Use other stochastic distances and test the NLM-Statistical filter on several images. To include the distances in the available codes is easy: just define a new function containing its mathematical description and call it from the core of the NLM-Statistical filter.

9. In the experiments shown, the simulated data (the *pattern*) has 4 classes with a gray level for each one. Other values can be selected to make the classification harder just by using gray level values closer for all classes. For instance, 20, 30, 40, 50 or other combinations. Besides, strong scatterers can be included or/and fine details, such a fine long lines. It is recommended to build a new more complicated pattern and to test some of the classification methods discussed in this chapter. Additionally, apart from the visual analysis, it is recommended to obtain the confusion matrices and the κ metric and to extract conclusions.

10. Select some other features for the case of the simulated data and compare results. Haralick's texture descriptors can be used. Confusion matrices and κ metrics are required also to have an objective assessment.

11 Select some pair of images from the SEN1-2 dataset, select some data and build a ground truth. Then, run some classification methods and obtain the κ metrics for actual SAR data.

12 By using the κ metric, it is interesting to ensemble a Monte Carlo experiment for simulated SAR data by doing this,

(a) Build the pattern (150×150), with some gray level values for each class (ground truth),
(b) select a number of Monte Carlo experiments (for instance, 1000 runs),
(c) select a level of speckle content (for example, ENL $= 1$) and simulate the data (generate the speckle from the *Gamma* distribution and multiply the values pixelwise with the ground truth data),
(d) run at least two classification methods for each of the generated patterns,
(e) estimate the κ metric for each pattern and each classification method.

Plot the results and extract conclusions.

13 By using the κ metric and the simulated SAR data, estimate the best number of neighbors for the K-nn method. That is, run several times (Monte Carlo) the K-nn method on several simulated data using $k = 1, k = 2, ...k = 5$ neighbors. It must be taken into account that a best κ metric does not imply always a best visual result.

14 Perform a classification analysis -using Monte Carlo- selecting one classification method (for instance, the K-means method) and several filters from the ones explained in this chapter (the Lee filter, the NLM-Statistical,...). Extract conclusions from the plots obtained.

6

Advanced Topics

This chapter deals with two challenging topics: filter assessment and robustness. The key concept regarding the assessment of a despeckling filter relies on the *metrics* used for the evaluation of its performances. Such metrics should also agree with human perception and should be easy to compute. Statistics applied to actual problems cannot be understood without discussing all relevant matters related to robustness, where robustness inference plays a fundamental role.

6.1 Assessment of Despeckling Filters

Assessment of despeckling filters is not a trivial issue. It resembles a multifaceted problem where many aspects must be considered. In this section, a set of well-established image-quality indices (metrics) are first introduced. Then, new advanced metrics are discussed.

6.2 Standard Metrics

There are some image-quality indices *inherited* from image processing for natural images, such as the well-known PSNR (Peak Signal-to-Noise Ratio) used in the evaluation of some of the filters discussed in this chapter. Other inherited metric is the MSSIM (Mean Structural Similarity Index, but is not that popular (see below). (Argenti et al., 2013) discuss several measures of speckle filter quality. They can be categorized as with-reference indexes, i.e. when the "perfect" or "noiseless" image is available, and without-reference indexes (the most usual case in practice for SAR imagery).

From the many metrics, if a ground truth is not available, the recommended image-quality indices recommended:

Mean preservation: The mean value of the noisy image μ_N estimated within a large enough homogeneous area (same radar signature) is compared to the mean value for the denoised image, μ_D. A good filtering operation should preserve that value ($\mu_N \approx \mu_D$).

Variance reduction: The variance of the filtered image, var_D must be smaller than the variance of the noisy image (var_N), both measured in a large enough textureless area.

ENL (equivalent number of looks): ENL is among the simplest and most spread measure of quality of despeckling filters. It can be easily estimated, in textureless areas and intensity

SAR Image Analysis — A Computational Statistics Approach: With R Code, Data, and Applications, First Edition. Alejandro C. Frery, Jie Wu, and Luis Gomez.

Companion website: www.wiley.com/go/frery/sarimageanalysis

format, as the ratio of the squared sample mean to the sample variance, i.e. the reciprocal of the squared coefficient of variation. ENL is proportional to the signal-to-noise ratio, the higher the ENL is, the better the quality of the image is in terms of speckle reduction. Much care must be taken when interpreting ENL; it is well known that large ENL values are easily obtained just by overfiltering an image, which severely degrades details and gives the filtered image an undesirable blurred appearance. ENL can be better estimated by the same formula used in the NLM-statistical filter, which is reproduced herein: the maximum likelihood estimator of L is the solution of $\ln \widehat{L} - \psi(\widehat{L}) - \ln \widehat{\lambda} + n^{-1} \sum_{i=1}^{n} \ln Z_i = 0$.

If a ground truth is available, apart from the above metrics, it is recommended to employ at least one image-quality index accounting for the preservation of statistical measures after filtering, and one metric for evaluating edges and fine details preservation. The ones recommended are as follows:

SSIM (structural similarity index): This measures the similarity between the original and despeckled images with local statistics (mean, variance and covariance between the unfiltered and despeckled pixel values). SSIM comprises, in one formula, three image measures: the luminance, the contrast, and the structure. SSIM is bounded in $(-1,1)$, and a good similarity produces value close to 1. It is available in `R` through the package `SpatialPack` package; cf. https://search.r-project.org/CRAN/refmans/SpatialPack/html/SSIM.html, and in Matlab with the function `ssimval`. Additionally, it is to easy implement through its mathematical definition:

$$\mathrm{SSIM}(x,y) = \frac{\left(2\widehat{\mu}_x\widehat{\mu}_y + c_1\right)\left(2\widehat{\sigma}_{xy} + c_2\right)}{\left(\widehat{\mu}_x^2 + \widehat{\mu}_y^2 + c_1\right)\left(\widehat{\sigma}_x^2 + \widehat{\sigma}_y^2 + c_2\right)}, \tag{6.1}$$

where x and y are selected areas of size $n \times m$ within the image, and $\widehat{\mu}_x$, $\widehat{\mu}_y$, $\widehat{\sigma}_x^2$, $\widehat{\sigma}_y^2$ are the sample mean and the sample variance estimated in the areas x and y, respectively, and $\widehat{\sigma}_{xy}$ is the covariance of x and y. The variables $c_1 = (k_1, D)^2$ and $c_2 = (k_2, D)^2$, where D is the dynamic range of the image[1], and $k_1 = 0.01$ and $k_2 = 0.03$ avoid division by zero. SSIM can be measured in the whole image or in several selected areas (x_i, y_i) and then averaged to get the MSSIM metric. For SAR images, it is recommended to estimate MSSIM.

β: Beta assesses edge preservation by measuring the correlation between edges in the reference image and the despeckled image (Achim et al., 2006). Edges are detected either by the Laplacian or the Canny filter. Both filters are available in Matlab by the method `edge`, and in `R` with the package `image.CannyEdges` (https://cran.r-project.org/web/packages/image.CannyEdges/image.CannyEdges.pdf). The β index ranges between 0 and 1, and the bigger it is, the better the filter is (the ideal edge preservation yields $\beta = 1$). Once the edges of both images, the original noisy, I and the despeckled one, $\widehat{I}$, are available, β is obtained through

$$\beta = \frac{\Gamma(\Delta I - \overline{\Delta I}, \widehat{\Delta I} - \overline{\widehat{\Delta I}})}{\sqrt{\Gamma(\Delta I - \overline{\Delta I}, \Delta I - \overline{\Delta I})}\sqrt{\Gamma(\widehat{\Delta I} - \overline{\widehat{\Delta I}}, \widehat{\Delta I} - \overline{\widehat{\Delta I}})}}, \tag{6.2}$$

where $\Gamma(I_1, I_2)$ is given by

$$\Gamma\left(I_1, I_2\right) = \sum_{i=1}^{K} I_{1_i} \cdot I_{2_i}. \tag{6.3}$$

1 For an 8 bit resolution image, i.e. an image with 255 possible grayscale values, $D = 2^8 - 1 = 255$.

ΔI and $\widehat{\Delta I}$ are the high-pass filtered versions of images I and $\hat{I}$, respectively, obtained with a sliding Laplacian pixel kernel window of size 3×3 or another edge detector such as the Canny detector; $\overline{\Delta I}$ and $\overline{\widehat{\Delta I}}$ are the average values of the image I and the average of the high-pass filtered version of the image $\widehat{\Delta I}$, respectively. This metric evaluates the correlation between the ground truth edges within the original image and the edges in the denoised image detected by means of the Laplacian filter (or the Canny filter).

SSIM and β are widely applied in SAR images and natural images, and a reference image is needed for their evaluation. The other three metrics (mean preservation, variance reduction, and ENL) do not need such a reference image.

Apart from the well-known metrics discussed above, the evaluation of the performances is completed by a visual inspection (by an expert) of the filtered data to assess the preservation of edges and fine details. Special care must be taken to assure that strong bright scatterers are well preserved, and that filters do not add artifacts to the final results.

6.2.1 Advanced Metrics for SAR Despeckling Assessment

To overcome the problems related to the unavailability of ground truth for most SAR data, new metrics have emerged recently that operate not on the filtered result but on the ratio image.

Some non-referenced metrics are available for image processing for natural images. Among them, BRISQUE (Blind/Referenceless Image Spatial QUality Evaluator) is widely applied. BRISQUE (Mittal et al., 2012) uses a model to address most common image distortions in natural images such as blur, Gaussian noise, artifacts due to compression operations, ringing, ... BRISQUE correlates well with human perception. However, BRISQUE has not been designed to deal with the particularities of SAR data, which are far from the shown by natural images. Assuming the multiplicative model, the observed image Z is the product of two independent fields: the backscatter X and the speckle Y. The observed data are $Z = XY$. An ideal estimator would be the one for which the ratio of the observed image to the filtered one $I = Z/\widehat{X}$ is only speckle: a collection of independent identically distributed samples from Gamma variates with unitary mean. Then, the quality of a filter is assessed by the closeness of I to the hypothesis that it is adherent to the statistical properties of pure speckle. An example of ratio images after applying two despeckling filters, namely the improved Lee filter (Lee et al., 2009) and the FANS filter (Cozzolino et al., 2014) is shown in Figure 6.1. A ratio image more similar to a random pattern (see Figure 6.2), that is, with no remains of geometric structure within is the ideal result.

Not only the visual inspection of the ratio images is required but also its numerical (*objective*) evaluation. This is addressed by Gomez et al. (2017b). Before this work, ratio images were only evaluated visually and numerically through its radiometric SAR signature (Achim et al., 2006), but no metrics for the geometric remains were available. That is, in a large enough selected area within the ratio image (preferable an area without geometric remains), the mean value should be 1 and the ENL value should be close to the nominal number of looks of the noisy image. This evaluation is shown in Figure 6.3 for the simulated SAR data used in the experiments. The simulated data (3 looks) is well despeckled. The measures within the ratio images are as expected: the mean value is close to 1, the ENL is close to the original one (3 looks). In addition, no visible geometric structure can be seen in the ratio image.

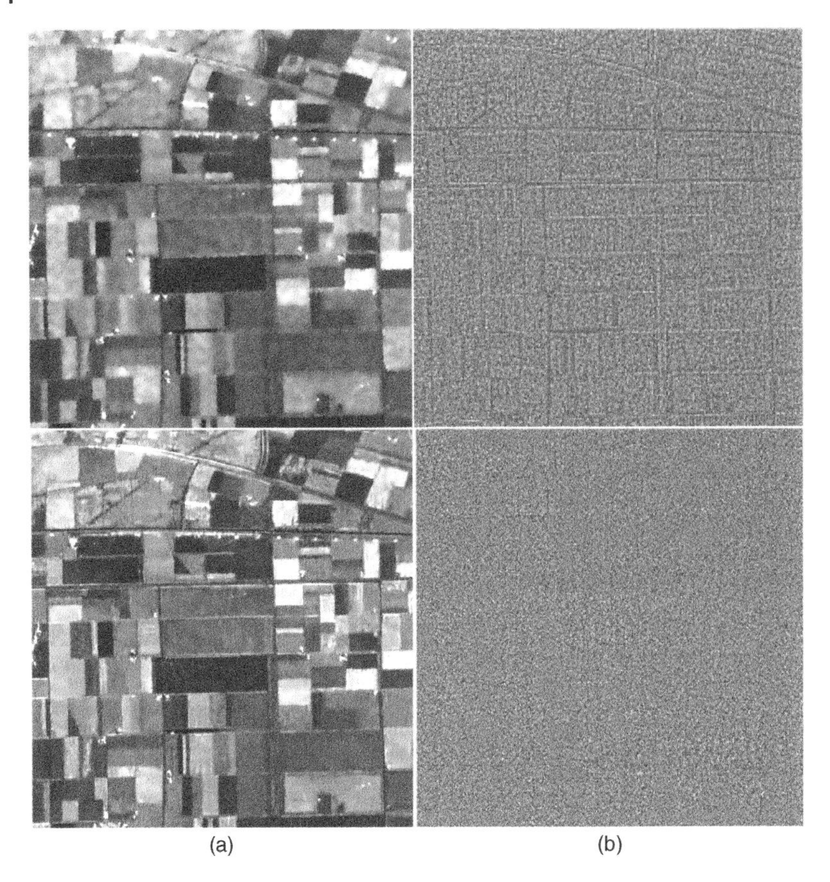

Figure 6.1 Examples for ratio images for the Flevoland SAR data. Top to bottom, (a) results of applying the E-Lee, and the FANS filters. Top to bottom (b), their ratio images.

Figure 6.2 Ideal ratio image.

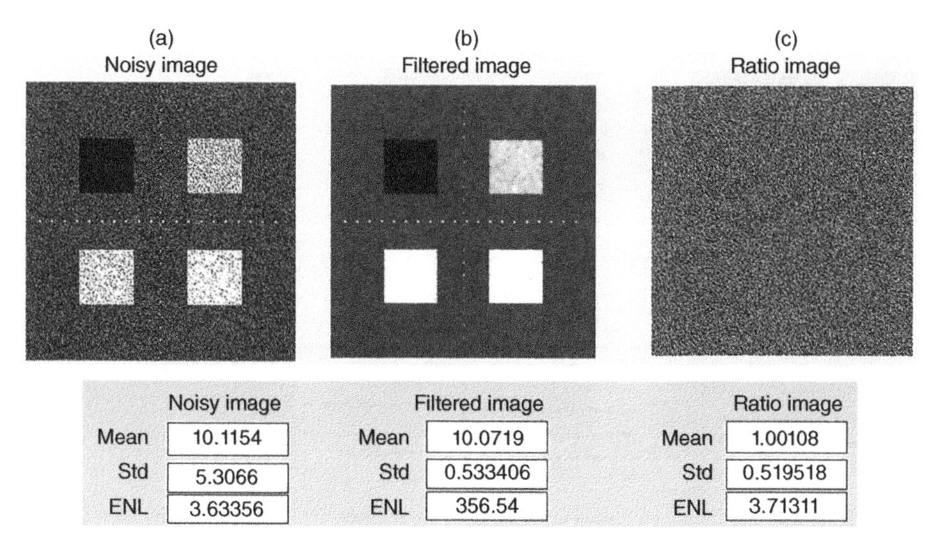

Figure 6.3 (a) Original 3 looks simulated SAR data and, the despeckled one (b) and, the ratio image (c). The metrics shown below have been estimated within a selected area (not shown in the figures).

To measure the geometric content in the ratio image is not trivial, especially for state-of-the-art despeckling filters where improvements are almost incremental and ratio images show little remains. Such little content is, sometimes, hardly visible noticeable. In Gomez et al. (2017b), a first attempt to get an objective evaluation of the filtered result not from the filtered image (the usual approach) but from the ratio image is researched. What is done in that work is to combine in a metric, a measure of the radiometric SAR signature with a numerical measure of the geometric contain; all estimated within the ratio image. Indeed, the proposed estimator, $\mathcal{M}$ is calculated as an average of n estimates as follows,

$$\mathcal{M} = r_{\widehat{\mathrm{ENL},\hat{\mu}}} + \delta h, \tag{6.4}$$

where the first element accounts for measuring the SAR radiometric signature, is calculated as

$$r_{\widehat{\mathrm{ENL},\hat{\mu}}} = \frac{1}{n}\sum_{i=1}^{n}\left(\frac{r_{\widehat{\mathrm{ENL}}}(i) + r_{\hat{\mu}}(i)}{2}\right), \tag{6.5}$$

where, for each homogeneous area i selected within the ratio image,

$$r_{\widehat{\mathrm{ENL}}}(i) = \frac{\left|\widehat{\mathrm{ENL}}_{\mathrm{noisy}}(i) - \widehat{\mathrm{ENL}}_{\mathrm{ratio}}(i)\right|}{\widehat{\mathrm{ENL}}_{\mathrm{noisy}}(i)}, \tag{6.6}$$

which is the absolute value of the relative residual due to deviations from the ideal *ENL*, and

$$r_{\hat{\mu}}(i) = \left|1 - \hat{\mu}_{\mathrm{ratio}}(i)\right|, \tag{6.7}$$

is the absolute value of the relative residual due to deviations from the ideal mean (which is 1). The ideal despeckling operation would yield $r_{\widehat{\mathrm{ENL},\hat{\mu}}} = 0$.

The Matlab script to get the $r_{\widehat{\mathrm{ENL},\hat{\mu}}}$ estimator is located in file

- `Code/Matlab/Chapter6/r_ENL_mu.m`.

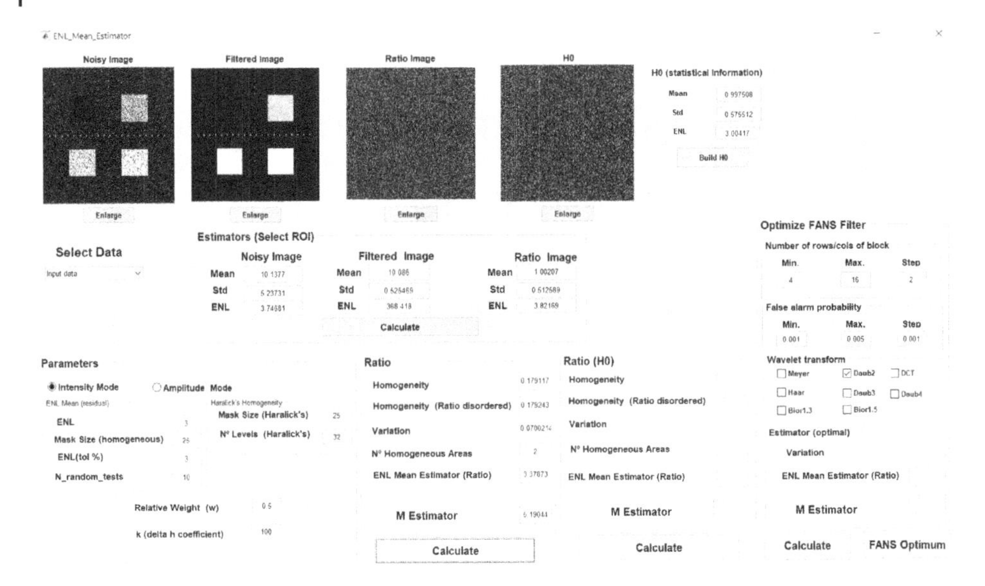

Figure 6.4 Matlab graphic interface to calculate the $\mathcal{M}$ metric (the link to the codes is provided in the text).

The second component of (6.4), δh, estimates the geometric content and it is calculated through measuring the homogeneity within the ratio image as follows,

$$h = \sum_i \sum_j \frac{1}{1 + (i-j)^2 \cdot p(i,j)}, \tag{6.8}$$

which corresponds to the inverse difference moment (also called homogeneity) from co-occurrence matrices (Haralick et al., 1973), where $p(i,j)$ is one value of such matrices at an arbitrary i,, j, position. Low values are associated with low textural variations and viceversa.

Therefore, from n areas selected within the ratio image, the $\mathcal{M}$ estimator is obtained and its value is compared with the $\mathcal{M}$ measure on a ratio image without any structure and ideal radiometric signature values. This last calculus is indeed not needed because for a collection of independent identically distributed samples without structure at all, $\mathcal{M} = 0$. The perfect despeckling filter will produce $\mathcal{M} = 0$, and the larger $\mathcal{M}$ is, the further the filter is from the ideal.

It is interesting to note that the $\mathcal{M}$ metric can be enriched with more elaborated statistical measures to capture deviations from the ideal *collection of independent identically distributed samples* in the ratio image. This must be done for evaluating state-of-the art despeckling filters and as mentioned above, the $\mathcal{M}$ is a first attempt.

Although easy to code, a link to an implementation in Matlab is provided in `Code/Matlab/Chapter7/M_estimator.m`. The code includes a friendly interface as depicted in Figure 6.4.

6.2.2 Completing the Assessment

The assessment is better done through a Monte Carlo analysis and producing results with statistical significance (Moschetti et al., 2006). A recommended setup is to run 1000 independent runs on simulated SAR data and on actual SAR data. Due to the potential lack of such huge volume of SAR data, an excellent approach is to select small areas of some SAR images and to perform the Monte Carlo analysis on them. Monte Carlo analysis should be also used to tune the filter parameters. We recommend the work by Martino et al. (2014) for and enriched filter assessment through large datasets and protocols (benchmarks). From the same Authors, the website http://www.grip.unina.it/research/80-sar-despeckling/85.html/ provides a freely available rich collection of simulated SAR data which includes strong bright scatterers.

If the purpose of the research or studio is to present a new despeckling filter, the recommended final step is to perform a sensitivity analysis that shows how the setting of the filter parameters (*tuning*) influences on the filtered results. A standard sensitivity analysis consists of varying a filter parameter over a particular range of values (for instance, the patch size, $p_s \in \{p_{s1}, p_{s1}, \dots, p_{sn}\}$) and see how this affects to the result. This must be done also through a Monte Carlo analysis and for all the filter parameters involved. The sensitivity analysis strongly contributes to assess the robustness of the method. Robustness (and robustness inference) is the topic of Section 6.3.

6.3 Robustness

We often see the term "robust" in the scientific literature as an adjective used to denote a desirable ability of models and procedure. Quoting Huber and Ronchetti (2009):

> [...] "robust" has now become a magic word, which is invoked in order to add respectability.

Robust inference should always be part of any statistical analysis, but it is of paramount importance in image processing. Consider the example of filters which employ data from small windows, as all those described in this book. Many statistical analyses are performed on these observations, often assuming an underlying model of independent and identically distributed random variables. This hypothesis fails when the observation window includes data from more than a single class, and such a failure may compromise the inference.

This section intends to clarify the concept of statistical robustness in the context of SAR data analysis. It is by no means a complete account of the literature. Rather than that, it aims at providing examples of what is needed to fairly use the term "robust" when attached to inference or to a procedure that depends on estimation. We provide examples and hints to develop robust inference techniques.

6.3.1 Robust Inference

Statistical inference uses two elements:

1. part of the observations (the whole set of observations is the *population*, which is often inaccessible), and
2. previous knowledge (*hypotheses*) about the population.

More often than not, along this book we have assumed that "the observations are independent and identically distributed according to $\mathcal{D}(\theta)$, with $\mathcal{D}$ a distribution indexed by the parameter $\theta \in \Theta \subset \mathbb{R}^p$." With this hypothesis, we have devised estimators, say by maximum

Listing 6.1 The "typing experiment"

```
require(statip)

N <- 100 # Sample size
R <- 10000 # Number of replications
epsilon <- 0.005 # Probability of contamination
cont.scale <- 100 # Scale of contamination

v.mean.pure <- rep(0, R)
v.median.pure <- rep(0, R)
v.mean.cont <- rep(0, R)
v.median.cont <- rep(0, R)

set.seed(1234567890, kind="Mersenne-Twister")
for(r in 1:R) {
        x <- rnorm(N)
        x.cont <- x
        ind.cont <- rbern(N, epsilon)
        x.cont <- x * (1-ind.cont) + x * ind.cont * cont.scale

        v.mean.pure[r] <- mean(x)
        v.mean.cont[r] <- mean(x.cont)
        v.median.pure[r] <- median(x)
        v.median.cont[r] <- median(x.cont)
}
```

likelihood $\hat{\theta}_{\text{ML}}$ or by analogy $\hat{\theta}_{\text{A}}$. But, what happens when the reality does not obey our hypothesis? The theory of quantitative robustness deals with this problem, and looks for estimators that are little affected by small deviations from the starting underlying assumptions. Such robustness comes at a price: if the data obey the model, the new estimator will not be optimal. A good robust estimator must, then, be "not too bad" under the main hypotheses, and "very good" under mild deviations from the assumed model.

6.3.2 The Mean and the Median

We will see the effect of a few outliers on the mean. Notice that we do not assume any statistical model.

Consider the situation of estimating the central location of the sample $\boldsymbol{x} = (x_1, x_2, \dots, x_{100})$. The obvious estimate is the sample mean $\overline{\boldsymbol{x}} = 100^{-1}\sum_j = 1^{100}x_1$. Assume further that the observations are listed with its integer part, tenths and hundredths, and that the person who is typing them in the computer is prone to missing the decimal point. We may describe this tendency to err as a random variable which occurs with probability $\epsilon = 5/1000$ (there is no fatigue or improvement in the process), and that the observations are typed independently. We will call this error "contamination." In order to be on the safe side, we will also use the median as an estimation of the central location. What can we say about these two estimators?

We devised a Monte Carlo experiment to analyze the behavior of the mean and the median in this "typing experiment." Listing 6.1 provides the core code for this experiment.

Line 1 loads the library that provides function `rbern` for sampling from the Bernoulli distribution. Lines 3 to 11 set up the constants and vectors used in the simulation. Line 13 sets the pseudorandom number generator seed and algorithm. The replication loops starts in line 14. Line 15 produces a sample from the standard Normal distribution (this is an arbitrary choice, as we did not assume any model). Line 17 samples from the Bernoulli distribution: "`0`" is correct typing, and "`1`" is an occurrence of error (missing the decimal point). We build the contaminated sample with this binary vector in line 18. Lines 20 to 23 compute the two

Table 6.1 Summary statistics from the "typing experiment."

	v.mean.pure	v.mean.cont	v.median.pure	v.median.cont
Min.	−0.385	−4.81	−0.501	−0.501
1st Qu.	−0.070	−0.12	−0.084	−0.084
Median	−0.003	0.00	−0.002	−0.002
Mean	−0.002	0.00	−0.002	−0.002
3rd Qu.	0.065	0.12	0.080	0.080
Max.	0.362	4.16	0.492	0.492

estimators with the pure and contaminated samples, and store the estimates for further analysis. Table 6.1 presents the summary statistics.

Table 6.1 reveals that all estimators in mean and median are close to the correct value: zero. The table also shows similar behavior of the mean for uncontaminated data (`v.mean.pure`), and the median for both uncontaminated (`v.median.pure`) and contaminated (`v.median.cont`) observations regarding the spread of the data, but a more spread when it comes to the mean with contaminated data (`v.mean.cont`). The minimum and maximum values of `v.mean.cont` are one order of magnitude larger than those of the other estimators; the first and third quartile are also scaled, approximately by a factor 2. In fact, notice that the summary statistics of the median applied to both pure and contaminated data are identical, as if the median had "deleted" the errors. As the only difference is contamination, we may conclude that the mean is susceptible to this type of contamination, whereas the median is not.

Figure 6.5 shows the histograms of the estimates from the above experiment. These histograms confirm that the central value is not affected by the contamination, but also that the errors introduce a noticeable spread only in the mean when applied to contaminated data.

Figure 6.6 shows the boxplots of the estimates from the above experiment. The boxplot of the mean when applied to contaminated data clearly shows the effect of errors with an increased number of outlying observations and a larger inter-quartile range.

Although the previous analysis sheds light on the distribution of the estimators under two conditions, it does not illuminate the specific impact of contamination on an individual trial. We will now see the effect of a single observation in the mean and the median.

Consider the situation of having a sample of thirty observations $\boldsymbol{x}' = (x_1, x_2, \dots, x30)$, and adding another observation, x, that will vary. We obtain a new sample of size 31 as $\boldsymbol{x} = (x, x_1, x_2, \dots, x_{30})$. The mean and the median of $\boldsymbol{x}$ are now functions of x, and we see how the vary in Figure 6.7.

The thirty observations from $\boldsymbol{x}'$ are shown as dots along the abscissas axis. The mean $\overline{\boldsymbol{x}'}$ is the solid highlighted horizontal line, and the median $q_{1/2}(\boldsymbol{x}')$ appears as the dashed highlighted horizontal line. The estimates do not coincide, as expected in general, but they are close. Moving $x \in [-5,5]$ renders the sample mean shown in solid line, which varies continuously and boundlessly. The median, on the contrary, varies within bounds imposed by the other observations. This behavior illustrates the limited effect that a single observation has on the median, a desirable property when the hypotheses may be compromised.

It is noteworthy that the median resists up to 50 % of contaminated observations before rendering a useless result. This proportion is known as *breakdown point*, and the median has

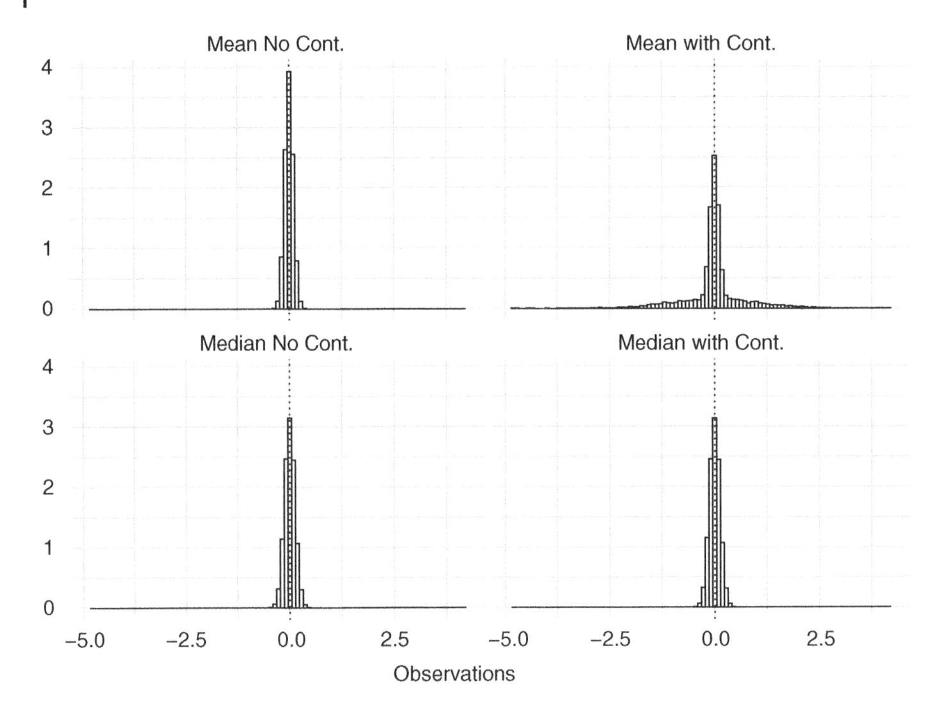

Figure 6.5 Histogram of 10 000 replications of the "typing experiment" with and without contamination.

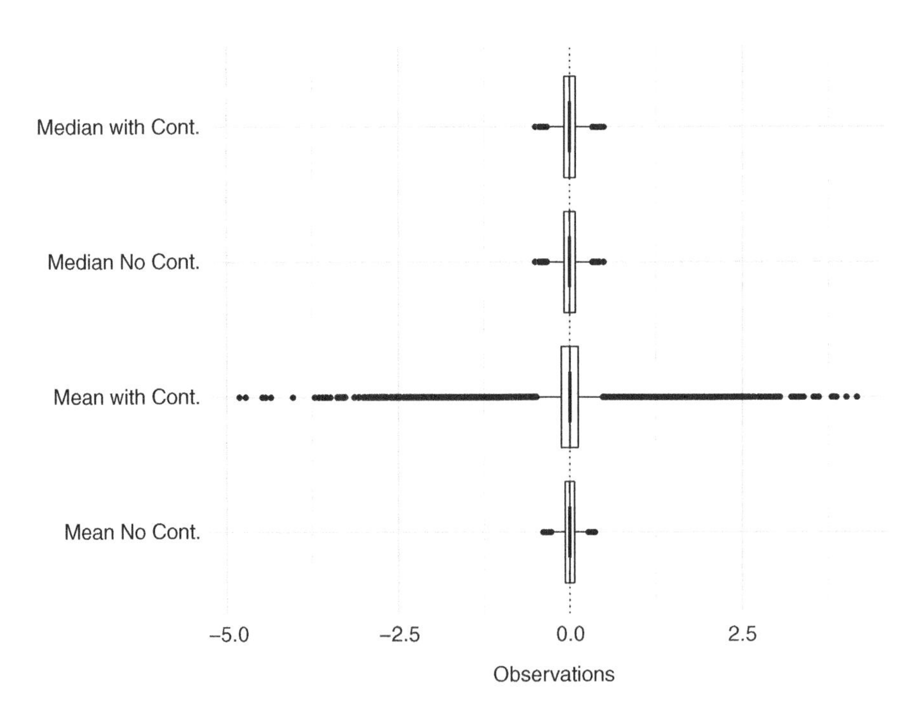

Figure 6.6 Boxplots of 10 000 replications of the "typing experiment" with and without contamination.

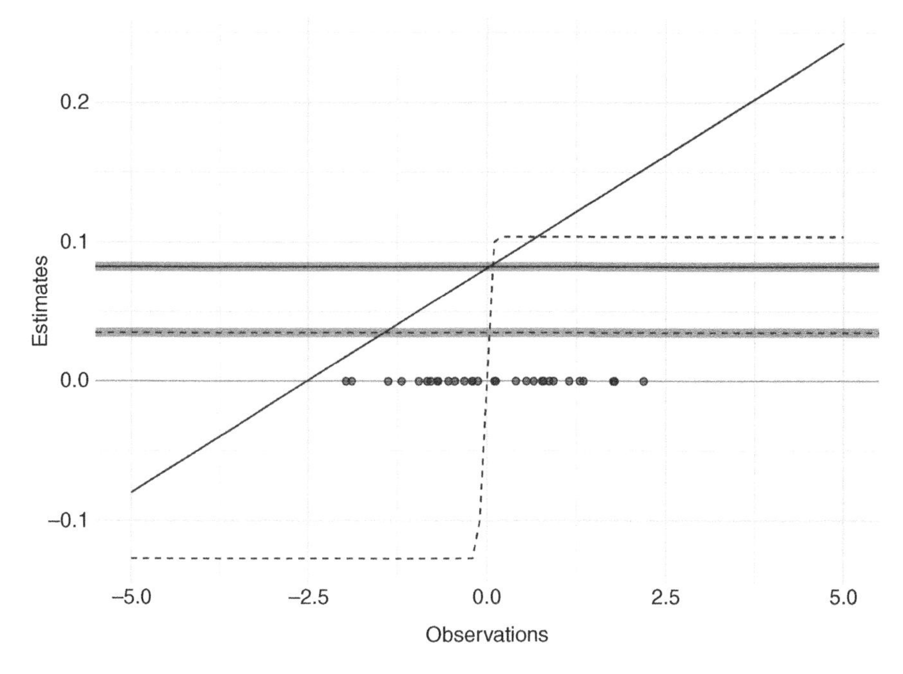

Figure 6.7 The influence on the mean and the median of moving a single observation in a sample.

the highest, while the mean has the lowest: 0 %, since a single observation can make it return an arbitrarily large value.

6.3.3 Empirical Stylized Influence Function

In this section we will advance the use of robustness twofold, namely: (i) by using a model, and (ii) by introducing influence functions. The main reference for this approach is the book by Hampel et al. (1986).

Assume that we have selected a sample $\boldsymbol{z}$ from an area with fully developed speckle. We may, then, use the SAR Gamma model described in Section 4.3.4 and estimate L by $\widehat{L}$, the equivalent number of looks. We will study three estimation techniques. The simplest is the inverse of the square of the coefficient of variation:

$$\widehat{L}_{\text{CV}} = \frac{\overline{\boldsymbol{z}}}{s^2(\boldsymbol{z})}. \tag{6.9}$$

We may also use our knowledge of the parametric model, and maximize the likelihood. This approach leads to the estimator

$$\log\left(\widehat{L}_{\text{ML}}\right) - \psi^{(0)}\left(\widehat{L}_{\text{ML}}\right) - \log(\overline{z}) + \frac{1}{n}\sum_{i=1}^{n} \log z_i = 0. \tag{6.10}$$

Although (6.9) is widely used, we have already seen that the sample mean is far from robust. We may, thus, replace the mean by the median and the variance by the square of the MAD (Median Absolute Deviation):

$$\widehat{L}_{\text{Rob}}(\boldsymbol{z}) = \left(\frac{q_{1/2}(\boldsymbol{z})}{\text{MAD}(\boldsymbol{z})}\right)^2, \tag{6.11}$$

where

$$\mathrm{MAD}(\mathbf{z}) = q_{1/2}\left(|z_1 - q_{1/2}(\mathbf{z})|, |z_2 - q_{1/2}(\mathbf{z})|, \dots, |z_n - q_{1/2}(\mathbf{z})|\right).$$

We want to assess the impact a single observation z in the sample $\mathbf{z} = (z, z_1, z_2, z_{n-1})$ has on these estimators. In order to simplify our presentation, assume further that the mean parameter μ is set to one.

Analogously to what was presented before, we may study the behavior of $\widehat{\theta}$ when z in $\mathbf{z}$ varies. In order to make the study independent of the particular values $\mathbf{z}' = (z_1, z_2, \dots, z_{n-1})$, we may use an "stylized sample" which consists of "typical" observations. Andrews et al. (1972) propose using the i-th quantile of the underlying distribution:

$$z_i = F^{-1}\left(\frac{i - 1/3}{n + 1/3}\right). \tag{6.12}$$

Figure 6.8 shows the Stylized Empirical Influence Functions (SEIF) of the estimators based on the coefficient of variation (6.9), on maximum likelihood (6.10), and on the robust version of the coefficient of variation. We show five sample sizes for each SEIF, corresponding to usual small windows: 9, 25, 49, 81, and 121.

We notice that the larger the sample is, the smaller the effect a single observation has on the estimation. This is consistent with our intuition of the importance of using large sample sizes whenever possible. The estimator based on the coefficient of variation $\widehat{L}_{\mathrm{CV}}$ seems more affected by large values than the one that maximizes the likelihood $\widehat{L}_{\mathrm{ML}}$, but the latter is sensitive to deviations in the other direction: it is sensitive to small values. This is probably due to the fact that $\widehat{L}_{\mathrm{ML}}$ involves logarithms, while $\widehat{L}_{\mathrm{CV}}$ does not. The effect of a single observation on the estimator based on a robust version of the coefficient of variation $\widehat{L}_{\mathrm{Rob}}$ is bounded, differently from the other two. It is also noteworthy that, apart from the case $n = 9$, this effect is systematically smaller.

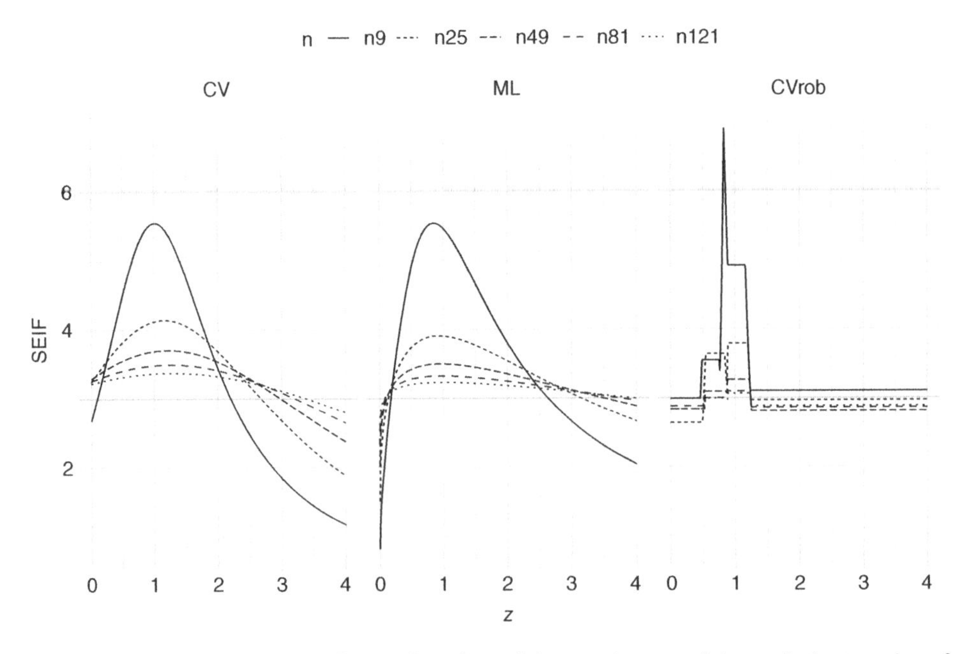

Figure 6.8 Stylized Empirical Influence Functions of three estimators of the equivalent number of looks and five sample sizes.

6.4 Rejoinder and Recommendations

In the first example we saw the influence that "wrong" observations have on the distribution of the mean and the median. We did not use any assumption about the data. The second example used a model, the SAR Gamma distribution, and proposed a robust version of a commonly used estimator for the equivalent number of looks. There are systematic approaches for obtaining robust estimators, among them the M- estimators that we briefly describe in the following.

M-estimators stem from a generalization of the maximum likelihood approach. Consider the sample $\boldsymbol{Z} = (Z_1, Z_2, \dots, Z_n)$ independent random variables that follow the same symmetric distribution $\mathcal{D}(\boldsymbol{\theta})$. For simplicity, let us consider the case with $\boldsymbol{\theta} = \theta \in \Theta \subset \mathbb{R}$. An M-estimator minimizes a global energy function

$$\widehat{\theta}_{\mathrm{M}} = \arg\min_{\theta\in\Theta} \mathfrak{E}(\theta)$$

defined in terms of a loss function ϱ:

$$\mathfrak{E}(\theta) = \sum_{i=1}^{n} \varrho(z_i; \theta). \tag{6.13}$$

Equivalently, we obtain the M-estimator by deriving (6.13) and equating to zero:

$$\sum_{i=1}^{n} \Psi(z_i; \theta) = 0, \tag{6.14}$$

where $\Psi(z_i; \theta) = \partial\varrho(z_i; \theta)/\partial\theta$.

The loss function ϱ is typically symmetric. Examples of such a function are $\varrho(z) = z^2/2$ (which yields the least squares estimator), and $\varrho(z) = |z|$ (which defines the median estimator). We obtain the maximum likelihood estimator setting $\varrho(z; \theta) = -\log f_Z(z; \theta)$, and with this choice $\Psi(z_i; \theta)$ is the score function. We achieve robustness by choosing $\Psi(z_i; \theta)$ as a composition of the score function and a bounded symmetric function, often defined piecewise.

We will mention three of these functions, namely those due to Huber, to Hampel and to Tukey.

$$\Psi_{\mathrm{Huber}}(z; b) = \max\{-b, \min\{b, z\}\}; \quad \text{with } 0 < b < \infty. \tag{6.15}$$

$$\Psi_{\mathrm{Hampel}}(z; a, b, r) = \begin{cases} z & \text{if } 0 \le |z| \le a, \\ a\,\mathrm{sign}(z) & \text{if } a \le |z| \le b, \\ a\dfrac{r-|z|}{r-b} & \text{if } b \le |z| \le r, \\ 0 & \text{if } r \le z; \quad \text{with } 0 < a \le b < r < \infty. \end{cases} \tag{6.16}$$

$$\Psi_{\mathrm{Tukey}}(z; k) = \begin{cases} z\left(1 - \left(\frac{z}{k}\right)^2\right)^2 & \text{if } |z| \le k < \infty, \\ 0 & \text{otherwise}; \quad \text{with } k > 0. \end{cases} \tag{6.17}$$

Figure 6.9 illustrates these functions (a: Huber, b: Hampel, and c: Tukey). The book by Hampel et al. (1986) is an authoritative reference about estimators based on influences

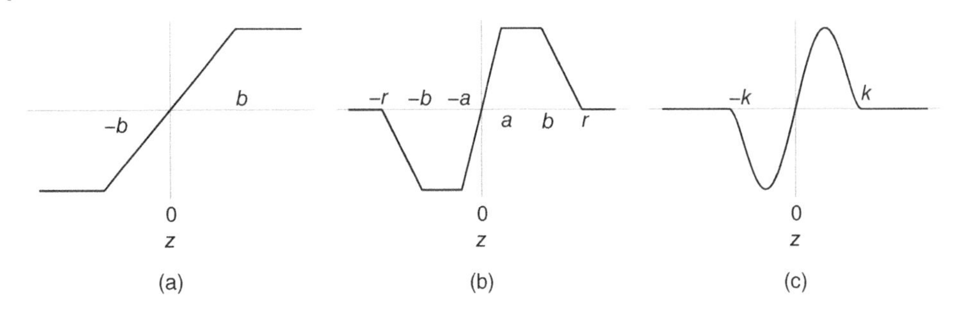

Figure 6.9 Bounded influence functions. (a) Huber, (b) Hampel, and (c) Tukey.

functions. This work also discusses how to choose such functions according to the expected (and, to some extent, unexpected) data behavior.

Once one chose a suitable influence function, there remains the question of which parameters are adequate for the problem at hand, e.g. b in (6.15), a, b and r in (6.16), and k in (6.17). Those constants must grant that the estimator is asymptotically unbiased and that it does not lose too much efficiency when there are no outliers.

Bustos et al. (2002) devised M-estimators for the $\mathcal{G}_A^0$ distribution. Allende et al. (2006), noting that the influence of outliers is asymmetric, proposed a class of M-estimators with asymmetric influence functions. These estimators, as expected, surpass the classical maximum likelihood approach in the presence of outliers, at the cost of more operations.

The approach based on influence functions is not the only way of obtaining robust estimators. Gambini et al. (2015) used asymmetric kernels to describe the data as a smoothed histogram, and then compute a distance between this histogram and a model. The model closest to the histogram provides the estimates. These estimators proved their robustness when used with few data in the presence of strong return as, for instance, from double bounce and corner reflectors.

We conclude this section quoting Wainer (1976):

> The methods described herein as robust are not to be dogmatically followed. They are not, in any sense, the best choice for all situations; rather, they are reasonable ways to start.

7

Reproducibility and Replicability

This chapter deals with reproducibility and replicability. Both terms are frequently confused even among researchers, therefore, at first, it is important to provide a clear definition for each word. Then, the implications of such relevant concepts in Remote Sensing in general, and in particular to the purposes aimed in this book, are discussed. In the last section of this chapter, a list of recommendations and suggestions to adhere to the good practices for making good research in SAR is drawn out. It is important to notice that reproducibility and replicability should permeate the whole scientific life. As noted by Wilson and Botham (2021), addressing these issues enhances your proposal by adding rigor and credibility.

7.1 What Is Reproducibility?

From the caves to here. From the first telephone to the smartphone, from ..., to ... It is a fascinating story, and sure, it has not been an easy one. Science and technology have been researched and done by *others* in the past. Fortunately, such a story continues every single day. Much of the advance comes from ideas that are then translated to experiments. In occasions, new experiments generate new ideas. Some experiments succeeded and others failed. No matters, all of them are necessary to push the knowledge ahead. But, no effective progress is achieved if it is not communicated to the world: it must be known, it must be published, it must be patented, it must be divulged, it must be broadcasted. *Others* must know what is new and the new issue must be visible (*experienced*) more than once by the *others*.

Reproducibility is at the very core of experimental sciences. Whatever result has been obtained by a researcher must be obtained easily by another researcher at low cost, without tricks. Therefore, and getting into the definition (Frery et al., 2020):

> "reproducibility consists of allowing the whole community to reach the researcher's shoulders. Scientific work is reproducible if other researchers can obtain the data and the code and, effortlessly, obtain the same products (analyses and reports)."

The many times cited Isaac Newton's metaphor (*If I have seen further, it is by standing on the shoulders of Giants*) lies at the above definition of reproducibility. However, although reproducible research has many benefits, it has also many challenges. Those challenges are mainly related to software. This is in the case of Remote Sensing. In other areas (for instance, pharmacology), the role of software is played by patents and potential benefits involved.

SAR Image Analysis — A Computational Statistics Approach: With R Code, Data, and Applications, First Edition.
Alejandro C. Frery, Jie Wu, and Luis Gomez.

Companion website: www.wiley.com/go/frery/sarimageanalysis

7.2 What Is Replicability?

The definition is also taken from Frery et al. (2020):

> "Scientific work is replicable if it reported in such a way that other researchers can perform similar studies and arrive at compatible conclusions."

In summary, reproducibility and replicability (of research) refer to what, in common sense, is understood as "good practices," or good science and technology. It is related to the honest researcher, investigator, inventor or technician. In general, "good science" has been the rule followed by most past and present actors of this huge adventure called science and technology. Unfortunately, "bad science" also co-existed with the good practices. Some areas are more susceptible than others to be affected by such unscientific approach. Remote Sensing is not one of the areas where bad practices are frequent. Many examples of "bad science" have been seen in Life Sciences. Some journals (even prestigious ones as *Science* or *Vaccines*) have retracted already published articles due to confirmation of bad practices from their authors (see https://science.sciencemag.org/content/356/6340/812.1, and https://science.sciencemag.org/content/373/6551/147.full).

The Turing Way Community et al. (2019) summarize these two concepts with a matrix of reproducibility, shown in Table 7.1. According to the authors:

> The different dimensions of reproducible research described in the matrix above have the following definitions:
>
> **Reproducible:** A result is reproducible when the same analysis steps performed on the same dataset consistently produces the same answer.
>
> **Replicable:** A result is replicable when the same analysis performed on different datasets produces qualitatively similar answers.
>
> **Robust:** A result is robust when the same dataset is subjected to different analysis workflows to answer the same research question (for example one pipeline written in R and another written in Python) and a qualitatively similar or identical answer is produced. Robust results show that the work is not dependent on the specificities of the programming language chosen to perform the analysis.
>
> **Generalisable:** Combining replicable and robust findings allow us to form generalizable results. Note that running an analysis on a different software implementation and with a different dataset does not provide generalized results. There will be many more steps to know how well the work applies to all the different aspects of the research question. Generalization is an important step toward understanding that the result is not dependent on a particular dataset nor a particular version of the analysis pipeline.

Table 7.1 Matrix of reproducibility.

		Data	
		Same	Different
Code	**Same**	reproducible	replicable
	Different	robust	generalisable

From The Turing Way Community et al. (2019).

7.3 Reproducibility and Replicability: Benefits for the Remote Sensing Community

A good example of the benefits of reproducible and replicable science comes from Isaac Newton. We must not to forget that Newton himself invented and built the first reflecting telescope (known later as *Newtonian telescope*), patiently polishing the mirror with his hands. Then, Newton published the invention in the book *Opticks: or, A treatise of the reflections, refractions, inflexions and colors of light.*[1] Most telescopes (including the Hubble) are Newtonian-like (reflectors)

It seems clear from that example what reproducibility means for society, for progress and, also for the researcher: there is no doubt whom invented the reflector telescope.

The same can not be said for other great contribution from Isaac Newton.

The controversy about who invented the calculus (differential and integral calculus), if was Isaac Newton or Gottfried Wilhelm Leibniz, was a hot issue for many years in the past. Known as "the calculus controversy," the question emerged because Newton did not publish his method after some decades later, when Leibniz's method was already published and devised independently from Newton's. In time, both methods were shown to be equivalent and the merit is for both creators, although Newton paid a prize for it: his mathematical notation (the *fluxions*) was replaced by the one proposed by Leibniz (differentials). That was a story with a happy ending, with both creators finally credited for, and the whole world was granted with new powerful mathematical tools.

SAR, as a part of the Remote Sensing community, is a very active research area, with many people involved and huge data that needs to be processed, in many cases, daily, in a frenetic stream, almost always in real time. SAR requires efficient algorithms that must evolve quickly (*adaptively*), following the ever-increasing capabilities of the technology.

Much of the progress of SAR is aimed at the need to publish that permeates the present science and technology. Researchers need to publish and technology improves from that new information that is published in papers, conferences, books, multimedia resources, etc. The number of Journals has never been so large in the history of science. The number of papers, obviously, is proportional to that figure: it seems endless. However, as a natural equilibrium, the threat of plagiarism is more alive than ever and it is already known public list of not-recommended Journal for the suspicious of fomenting bad practices.

Although benefits for reproducibility and replicability for everyone are crystal-clear, it seems that the benefits of making bad science somehow compensate others.

7.4 Recommendations for Making "Good Science"

Discuss models, speckle, statistical distributions, etc. with clarity, soundly, in a dynamic and educational way. It was one of the priorities of the Authors of this book. But in the mind of all, reproducibility and replicability were as important as the mathematical notation used. And, it was not just triggered by the this book's building pillars: some publications and work done before gives some credit to what reproducibility and replicability means for this book's Authors Frery et al. (2020).

---

1 https://doi.org/10.5479/sil.302475.39088000644674

Nothing like an example to clear out: this book is fully reproducible and fully replicable because,

- Reproducible: All the data and codes to get the many academic examples contained in it are available to users in a website. Besides, all codes are scripts that can be run by a single click.
- Replicable: Codes are all extensively commented and linked to the theoretical explanations provided in this book.

Additionally, plagiarism has been always avoided: when any explanation was *standing of the shoulders of Giants*, the corresponding credit to the *Giants* was given with enormous pleasure.

From the experience got, in what follows, a list of recommendations to make good science is provided. Sure there are more, and even the readers have their own. To share them with others is a good practice!

The list of recommendations: general lines

- The first one is the trivial one; to make good science, one wishes to make good science. It requires to be aware of what reproducibility and replicability are all about. First, know; then act.
- To make good science is indeed like studying a new subject in the Faculty: there are known rules, there are books that help and also, websites, researchers involved in and, even scientific societies (IEEE is one of these societies strongly committed to good practices). Therefore, there is an established methodology that promotes these practices.

The list of recommendations: the methodology.[2]

- To avoid plagiarism always cite all source (papers, books, codes, websites, personal communications, etc.) used in the research. There is nothing wrong to be generous on this topic.
- Support all results with codes. If possible, use open software architectures to make accessible the codes and the data to everyone. Codes must be commented on and linked to the publication (book, paper, website, etc.)
- It is very important that codes can be easily executed and downloaded. All the resources required for running the codes must be provided (external libraries). This is of great relevance in SAR due to the need for many different formats to deal with data and, for example, produce final images (libraries to manage/generate "`.png`," "`.jpg`," etc.)
- All codes must use relative paths to libraries and data to be used. That is, the recommended structure is like, `../../any-path/MyFigures`, and `../../any-path/MyVideoDemos` or `../../any-path/MyPersonalLibrary`.
- To include a complete example avoids problems to the user.
- Additionally, the codes must be "clean" and the input parameters must be checked out before running the codes. Of course, all problems related to division by zero or similar cases that result in NaN (Not a Number) must be properly handled. Matrices dimension must be always under control to avoid unexpected and difficult to locate errors.

2 Most of these recommendations are from Frery et al. (2020)

- Special care must be taken when codes are ready to be run in parallel (for instance, by using the `parfor` or `foreach` resources to make loops in parallel) or supported by Graphics Processing Unit (GPU) resources: provide always a full sequential version whenever possible.
- Finally, `readme.txt` files must be included anywhere needed.

Since codes, data, results, and other resources will be allocated in websites (Github is an excellent option), a common recommendation is to assure that repositories are keep updated for at least some years (a minimum of five years seems to be currently accepted).

Regarding experiments, as many times pointed out in this book, Monte Carlo analysis must be the unique option to support experiments with a statistical significance. As also previously justified, as Monte Carlo analysis lies on random numbers, to guarantee the replicabilility of experiments, the seeds to generate that random series must be under control, and the algorithm stated. A strongly recommended practice to better support results shown in experiments is to run Monte Carlo not on *toy-experiments* but on public datasets (benchmarks). As recommended in the Nature article by Wilson and Botham (2021):

> "We will calibrate our new method using a landmark dataset, a gold-standard comparison in our field, to benchmark against [...]. A new method could bias results, but benchmarking the method against a well-regarded dataset of known effects justifies the method's adequacy..."

Such approach is well established in most of the Life Sciences, where strict protocols and standards for testing have been devised. This is not the case for other areas of research although some efforts recently have emerged (see, for instance, Martino et al., 2014, Gomez et al., 2016, Mao et al., 2020, Martino et al., 2022).

Most of these recommendations are already followed by some IEEE Journals and, for instance, by IPOL (Image Processing On Line https://www.ipol.im/) and also by many research groups that broadcast their papers and codes on personal websites.

Two additional issues are addressed,

- (Another trivial one) Reproducibility and replicability do not mean that one works hard *for free*. All material located in repositories must be protected by a license. Creative Commons https://creativecommons.org/licenses/by-nd/3.0/us/ offer creators specific royalty-free uses for their work. The protection covers the copy and distribution of the material in any medium or format. Another possibility comes from GNU General Public License (GPL) http://www.gnu.org/licenses/licenses.en.html#GPL which is commonly used nowadays by most GNU programs to protect software (both, source and compiled files). Generally, to be protected by these licenses, it is enough to include in all codes (all material published) headers like the one shown in Figure 7.1. But the owner of the material must read all the requirements.
- Deep Learning: Learning from massive database techniques are becoming the usual approach nowadays, with enormous success, in most areas of Remote Sensing. However, it is not clear how to make codes, data (huge training databases!) available for users and more problematic, how to reproduce data. Certainly, this is a key problem to be solved in next years; cf. the work by Hartley and Olsson (2020).

```
/*
 * Copyright 2020, 2021
 *
 * This program is free software: you can redistribute it and/or modify
 * it under the terms of the GNU General Public License as published by
 * the Free Software Foundation, either version 3 of the License, or
 * (at your option) any later version.
 *
 * This program is distributed in the hope that it will be useful,
 * but WITHOUT ANY WARRANTY; without even the implied warranty of
 * MERCHANTABILITY or FITNESS FOR A PARTICULAR PURPOSE.  See the
 * GNU General Public License for more details.
 *
 * You should have received a copy of the GNU General Public License
 * along with this program.  If not, see <http://www.gnu.org/licenses/>.
 */
```

Figure 7.1 Usual header to include in codes to be protected by the GNU license.

7.5 Conclusions

This chapter provides some clues to invite researchers to adhere to the good practices to make good science. This "good science," of course, must be understood as reproducible and replicable science[3].

From the several recommendations given above, the one more important is again repeated, this time and, to conclude this book,

To make good science, one wishes to make good science.

3 It seems there are no rules to assure good science in the other sense. At least, the Authors of this book are not aware of them!

Bibliography

M. Abramowitz and I. A. Stegun. *Handbook of Mathematical Functions: With Formulas, Graphs.* Dover, New York, 1964.

A. Achim, E. E. Kuruoğlu, and J. Zerubia. SAR image filtering based on the heavy-tailed Rayleigh model. *IEEE Transactions on Image Processing*, 15(9):2686–2693, 2006. doi: https://doi.org/10.1109/TIP.2006.877362.

H. Allende, A. C. Frery, J. Galbiati, and L. Pizarro. M-estimators with asymmetric influence functions: the GA0 distribution case. *Journal of Statistical Computation and Simulation*, 76(11):941–956, 2006. doi: https://doi.org/10.1080/10629360600569154.

D. Andrews, P. Bickel, F. Hampel, P. J. Huber, W. Rogers, and J. Tukey. *Robust Estimates of Location: Survey and Advances.* Princeton University Press, Princeton, NJ, 1972.

F. Argenti and L. Alparone. Speckle removal from SAR images in the undecimated wavelet domain. *IEEE Transactions on Geoscience and Remote Sensing*, 40(11):2363–2374, 2002. doi: https://doi.org/10.1109/TGRS.2002.805083.

F. Argenti, A. Lapini, T. Bianchi, and L. Alparone. A tutorial on speckle reduction in synthetic aperture radar images. *IEEE Geoscience and Remote Sensing Magazine*, 1(3):6–35, 2013. doi: https://doi.org/10.1109/MGRS.2013.2277512.

A. Baraldi and F. Parmiggiani. An alternative form of the Lee filter for speckle suppression in SAR images. *Graphical Models and Image Processing*, 57(1):75–78, 1995. doi: https://doi.org/10.1006/gmip.1995.1008.

G. E. Box, J. S. Hunter, and W. G. Hunter. *Statistics for Experimenters: Design, Discovery and Innovation.* Wiley, Hoboken, NJ, 2 edition, 2005.

A. Buades, B. Coll, and J.-M. Morel. A non-local algorithm for image denoising. In *2005 IEEE Computer Society Conference on Computer Vision and Pattern Recognition (CVPR'05)*, volume 2, pages 60–65. Institute of Electrical and Electronics Engineers (IEEE), June 2005a. doi: https://doi.org/10.1109/cvpr.2005.38.

A. Buades, B. Coll, and J. M. Morel. A review of image denoising algorithms, with a new one. *Multiscale Modeling and Simulation*, 4(2):490–530, 2005b. doi: https://doi.org/10.1137/040616024.

A. Buades, B. Coll, and J. M. Morel. Image denoising methods. A new nonlocal principle. *SIAM Review*, 52(1):113–147, 2010. ISSN 00361445. doi: https://doi.org/10.1137/090773908.

O. H. Bustos and A. C. Frery. Reporting Monte Carlo results in statistics: suggestions and an example. *Chilean Journal of Statistics*, 9(2):46–95, 1992. http://chjs.soche.cl/index.php?option=com_content&view=article&id=70.

SAR Image Analysis — A Computational Statistics Approach: With R Code, Data, and Applications, First Edition.
Alejandro C. Frery, Jie Wu, and Luis Gomez.

Companion website: www.wiley.com/go/frery/sarimageanalysis

O. H. Bustos, M. M. Lucini, and A. C. Frery. M-estimators of roughness and scale for GA0-modelled SAR imagery. *EURASIP Journal on Advances in Signal Processing*, 2002(1):105–114, 2002. doi: https://doi.org/10.1155/S1110865702000392.

D. Chan, A. Rey, J. Gambini, and A. C. Frery. Sampling from the GI0 distribution. *Monte Carlo Methods and Applications*, 24(4):271–287, 2018. doi: https://doi.org/10.1515/mcma-2018-2023.

D. Cozzolino, S. Parrilli, G. Scarpa, G. Poggi, and L. Verdoliva. Fast adaptive nonlocal SAR despeckling. *IEEE Geoscience and Remote Sensing Letters*, 11(2):524–528, 2014. doi: https://doi.org/10.1109/LGRS.2013.2271650.

F. Cribari-Neto, A. C. Frery, and M. F. Silva. Improved estimation of clutter properties in speckled imagery. *Computational Statistics and Data Analysis*, 40(4):801–824, 2002. doi: https://doi.org/10.1016/S0167-9473(02)00102-0.

C.-A. Deledalle, L. Denis, and F. Tupin. Iterative weighted maximum likelihood denoising with probabilistic patch-based weights. *IEEE Transactions on Image Processing*, 18(12):2661–2672, 2009. ISSN 1057-7149. doi: https://doi.org/10.1109/tip.2009.2029593.

C.-A. Deledalle, V. Duval, and J. Salmon. Non-local methods with shape-adaptive patches (NLM-SAP). *Journal of Mathematical Imaging and Vision*, 43(2):103–120, 2011. doi: https://doi.org/10.1007/s10851-011-0294-y.

Y. Delignon and W. Pieczynski. Modeling non-Rayleigh speckle distribution in SAR images. *IEEE Transactions on Geoscience and Remote Sensing*, 40(6):1430–1435, 2002. doi: https://doi.org/10.1109/TGRS.2002.800234.

L. Devroye. *Non-Uniform Random Variate Generation*. Springer-Verlag, New York, 1986.

G. Di Martino, M. Poderico, G. Poggi, D. Riccio, and L. Verdoliva. Benchmarking framework for SAR despeckling. *IEEE Transactions on Geoscience and Remote Sensing*, 52(3):1596–1615, 2014. doi: https://doi.org/10.1109/TGRS.2013.2252907.

C. Farley, A. E. Raftery, T. B. Murphy, and L. Strucca. mclust version 4 for R: normal mixture modeling for model-based clustering, classification, and density estimation. Technical Report 597, Department of Statistics, University of Washington, June 2012.

W. Feng, H. Lei, and Y. Gao. Speckle reduction via higher order total variation approach. *IEEE Transactions on Image Processing*, 23(4):1831–1843, 2014. doi: https://doi.org/10.1109/TIP.2014.2308432.

D. Freedman and P. Diaconis. On the histogram as a density estimator: L2 theory. *Zeitschrift für Wahrscheinlichkeitstheorie und Verwandte Gebiete*, 57(4):453–476, 1981. doi: https://doi.org/10.1007/bf01025868.

A. C. Frery, H.-J. Müller, C. C. F. Yanasse, and S. J. S. Sant'Anna. A model for extremely heterogeneous clutter. *IEEE Transactions on Geoscience and Remote Sensing*, 35(3):648–659, 1997. doi: https://doi.org/10.1109/36.581981.

A. C. Frery, F. Cribari-Neto, and M. O. Souza. Analysis of minute features in speckled imagery with maximum likelihood estimation. *EURASIP Journal on Advances in Signal Processing*, 2004(16):2476–2491, 2004. doi: https://doi.org/10.1155/S111086570440907X.

A. C. Frery, J. Jacobo-Berlles, J. Gambini, and M. Mejail. Polarimetric SAR image segmentation with B-Splines and a new statistical model. *Multidimensional Systems and Signal Processing*, 21:319–342, 2010. doi: https://doi.org/10.1007/s11045-010-0113-4.

A. C. Frery, A. D. C. Nascimento, and R. J. Cintra. Analytic expressions for stochastic distances between relaxed complex Wishart distributions. *IEEE Transactions on Geoscience and Remote Sensing*, 52(2):1213–1226, 2014. doi: https://doi.org/10.1109/TGRS.2013.2248737.

A. C. Frery, L. Gomez, and A. C. Medeiros. A badging system for reproducibility and replicability in remote sensing research. *IEEE Journal of Selected Topics on Applied Earth Observations and Remote Sensing*, 13:4988–4995, 2020. doi: https://doi.org/10.1109/JSTARS.2020.3019418.

O. Frey, M. Santoro, C. L. Werner, and U. Wegmuller. DEM-based SAR pixel-area estimation for enhanced geocoding refinement and radiometric normalization. *IEEE Geoscience and Remote Sensing Letters*, 10(1):48–52, 2013a. doi: https://doi.org/10.1109/LGRS.2012.2192093.

O. Frey, C. L. Werner, U. Wegmuller, A. Wiesmann, D. Henke, and C. Magnard. A car-borne SAR and InSAR experiment. In *IEEE International Geoscience and Remote Sensing Symposium – IGARSS*, pages 93–96, 2013b. doi: https://doi.org/10.1109/IGARSS.2013.6721100.

J. Gambini, J. Cassetti, M. M. Lucini, and A. C. Frery. Parameter estimation in SAR imagery using stochastic distances and asymmetric kernels. *IEEE Journal of Selected Topics in Applied Earth Observations and Remote Sensing*, 8(1):365–375, 2015. doi: https://doi.org/10.1109/JSTARS.2014.2346017.

L. Gomez, L. Alvarez, R. de Lima Pinheiro, and A. C. Frery. A benchmark for despeckling filters. In *EUSAR 2016: 11th European Conference on Synthetic Aperture Radar*, pages 1–4, 2016.

L. Gomez, L. Alvarez, L. Mazorra, and A. C. Frery. Fully PolSAR image classification using machine learning techniques and reaction-diffusion systems. *Neurocomputing*, 255:52–60, 2017a. ISSN 0925-2312. doi: https://doi.org/10.1016/j.neucom.2016.08.140. https://www.sciencedirect.com/science/article/pii/S0925231217305465. Bioinspired Intelligence for machine learning.

L. Gomez, R. Ospina, and A. C. Frery. Unassisted quantitative evaluation of despeckling filters. *Remote Sensing*, 9(9):389, 2017b. doi: https://doi.org/10.3390/rs9040389. https://github.com/Raydonal/UNASSISTED.

I. Goodfellow, Y. Bengio, and A. Courville. *Deep Learning (Adaptive Computation and Machine Learning series)*. MIT Press, Boston, MA, 2016.

J. W. Goodman. Some fundamental properties of speckle. *Journal of the Optical Society of America*, 66(11):1145–1150, 1976.

I. S. Gradshteyn and I. M. Ryzhik. *Tables of Integrals, Series and Products*. Academic Press, New York, 1980.

R. Grimson, N. Morandeira, and A. C. Frery. Comparison of nonlocal means despeckling based on stochastic measures. In *IEEE International Geoscience and Remote Sensing Symposium – IGARSS*, pages 3091–3094, 2015. doi: https://doi.org/10.1109/IGARSS.2015.7326470.

Q. Guo, C. Zhang, Y. Zhang, and H. Liu. An efficient SVD-based method for image denoising. *IEEE Transactions on Circuits and Systems for Video Technology*, 26(5):868–880, 2016. ISSN 1558-2205. doi: https://doi.org/10.1109/TCSVT.2015.2416631.

F. R. Hampel, E. M. Ronchetti, P. J. Rousseeuw, and W. A. Stahel. *Robust Statistics: The Approach Based on Influence Functions*. Wiley, New York, 1986.

R. M. Haralick, K. Shanmugam, and I. Dinstein. Textural features for image classification. *IEEE Transactions on Systems, Man, and Cybernetics*, SMC-3(6):610–621, 1973. doi: https://doi.org/10.1109/TSMC.1973.4309314.

M. Hartley and T. S. G. Olsson. dtoolAI: Reproducibility for deep learning. *Patterns*, 1(5):100073, 2020. doi: https://doi.org/10.1016/j.patter.2020.100073.

T. Hastie, R. Tibshirani, and J. Friedman. *The Elements of Statistical Learning: Data Mining, Inference, and Prediction*. Springer, New York, 2 edition, 2017.

A. Henningsen and O. Toomet. maxLik: A package for maximum likelihood estimation in R. *Computational Statistics*, 26(3):443–458, 2011. doi: https://doi.org/10.1007/s00180-010-0217-1.

P. J. Huber. *Robust Statistics*. Wiley, New York, 1981.

P. J. Huber and E. M. Ronchetti. *Robust Statistics*. Wiley, 2009.

R. J. Hyndman and Y. Fan. Sample quantiles in statistical packages. *The American Statistician*, 50(4):361, 1996. doi: https://doi.org/10.2307/2684934.

E. Jakeman and P. N. Pusey. A model for non-Rayleigh sea echo. *IEEE Transactions on Antennas and Propagation*, 24(6):806–814, 1976.

N. L. Johnson, S. Kotz, and A. W. Kemp. *Univariate Discrete Distributions*. Wiley Series in Probability and Mathematical Statistics. John Wiley & Sons, New York, 2 edition, 1993.

N. L. Johnson, S. Kotz, and N. Balakrishnan. *Continuous Univariate Distributions*, volume 2. John Wiley & Sons, New York, 2 edition, 1995.

C. Kervrann, J. Boulanger, and P. Coupé. Bayesian non-local means filter, image redundancy and adaptive dictionaries for noise removal. In *Scale Space and Variational Methods in Computer Vision*, pages 520–532. Springer-Verlag, Berlin, Heidelberg, 2007. ISBN 978-3-540-72823-8.

V. C. Koo and Y. K. Chan et al. A new unmanned aerial vehicle synthetic aperture radar for environmental monitoring. *Progress in Electromagnetics Research*, 122:245–268, 2012. doi: https://doi.org/doi:10.2528/PIER11092604.

D. T. Kuan, A. A. Sawchuk, T. C. Strand, and P. Chavel. Adaptive noise smoothing filter for images with signal-dependent noise. *IEEE Transactions on Pattern Analysis and Machine Intelligence*, PAMI-7(2):165–177, 1985. doi: https://doi.org/10.1109/TPAMI.1985.4767641.

J.-S. Lee. Digital image enhancement and noise filtering by use of local statistics. *IEEE Transactions on Pattern Analysis and Machine Intelligence*, PAMI-2(2):165–168, 1980. doi: https://doi.org/10.1109/TPAMI.1980.4766994.

J.-S. Lee and E. Pottier. *Polarimetric Radar Imaging: from basics to applications*. CRC Press, Taylor & Francis Group, Boca Raton, FL, 1 edition, 2009.

J.-S. Lee, L. Jurkevich, P. Dewaele, P. Wambacq, and A. Oosterlinck. Speckle filtering of synthetic aperture radar images: a review. *Remote Sensing Reviews*, 8(4):313–340, 1994. doi: https://doi.org/10.1080/02757259409532206.

J.-S. Lee, J.-H. Wen, T. L. Ainsworth, K.-S. Chen, and A. J. Chen. Improved sigma filter for speckle filtering of SAR imagery. *IEEE Transactions on Geoscience and Remote Sensing*, 47(1):202–213, 2009. doi: https://doi.org/10.1109/TGRS.2008.2002881.

G. Liu, H. Zhong, and L. Jiao. Comparing noisy patches for image denoising: a double noise similarity model. *IEEE Transactions on Image Processing*, 24(3):862–872, 2015. ISSN 1057-7149. doi: https://doi.org/10.1109/TIP.2014.2387390.

A. Lopes, R. Touzi, and E. Nezry. Adaptive speckle filters and scene heterogeneity. *IEEE Transacntions on Geoscience and Remote Sensing*, 28(6):992–1000, 1990. doi: https://doi.org/10.1109/36.62623.

X. Ma, C. Wang, Z. Yin, and P. Wu. SAR image despeckling by noisy reference-based deep learning method. *IEEE Transactions on Geoscience and Remote Sensing*, 58(12):8807–8818, 2020. doi: https://doi.org/10.1109/TGRS.2020.2990978.

M. Mahmoudi and G. Sapiro. Fast image and video denoising via nonlocal means of similar neighborhoods. *IEEE Signal Processing Letters*, 12(12):839–842, 2005. doi: https://doi.org/0.1109/LSP.2005.859509.

C. F. Manski. *Analog Estimation Methods in Econometrics*, volume 39 of Monographs on Statistics and Applied Probability. Chapman & Hall, New York, 1988. http://elsa.berkeley .edu/books/analog.html.

Y. Mao, X. Li, H. Su, Y. Zhou, and J. Li. Ship detection for SAR imagery based on deep learning: a Benchmark. In *IEEE 9th Joint International Information Technology and Artificial Intelligence Conference (ITAIC)*, pages 1934–1940, 2020. doi: https://doi.org/10.1109/ITAIC49862.2020.9339055.

R. A. Maronna, R. D. Martin, and V. J. Yohai. *Robust Statistics: Theory and Methods*. Wiley Series in Probability and Statistics. Wiley, England, 2006.

G. D. Martino, A. D. Simone, A. Iodice, and D. Riccio. Benchmarking framework for multitemporal SAR despeckling. *IEEE Transactions on Geoscience and Remote Sensing*, 60:1–26, 2022, Art no. 5207826. doi: https://doi.org/10.1109/TGRS.2021.3074435.

F. N. S. Medeiros, N. D. A. Mascarenhas, and L. F. Costa. Evaluation of speckle noise MAP filtering algorithms applied to SAR images. *International Journal of Remote Sensing*, 24(24):5197–5218, 2003. doi: https://doi.org/10.1080/0143116031000115148.

M. E. Mejail, A. C. Frery, J. Jacobo-Berlles, and O. H. Bustos. Approximation of distributions for SAR images: proposal, evaluation and practical consequences. *Latin American Applied Research*, 31:83–92, 2001. http://www.laar.uns.edu.ar/indexes/i31_02.htm.

M. E. Mejail, J. Jacobo-Berlles, A. C. Frery, and O. H. Bustos. Classification of SAR images using a general and tractable multiplicative model. *International Journal of Remote Sensing*, 24(18):3565–3582, 2003. doi: https://doi.org/10.1080/0143116021000053274.

A. Mittal, A. K. Moorthy, and A. C. Bovik. No-reference image quality assessment in the spatial domain. *IEEE Transactions on Image Processing*, 21(12):4695–4708, 2012. doi: https://doi.org/10.1109/TIP.2012.2214050.

A. Moreira. Improved multilook techniques applied to SAR and Scan SAR imagery. *IEEE Transactions on Geoscience and Remote Sensing*, 29(4):529–534, 1991.

E. Moschetti, M. G. Palacio, M. Picco, O. H. Bustos, and A. C. Frery. On the use of Lee's protocol for speckle-reducing techniques. *Latin American Applied Research*, 36(2):115–121, 2006.

J. Naranjo-Torres, J. Gambini, and A. C. Frery. The geodesic distance between GI0 models and its application to region discrimination. *IEEE Journal of Selected Topics in Applied Earth Observations and Remote Sensing*, 10(3):987–997, 2017. doi: https://doi.org/10.1109/JSTARS.2017.2647846.

A. D. C. Nascimento, R. J. Cintra, and A. C. Frery. Hypothesis testing in speckled data with stochastic distances. *IEEE Transactions on Geoscience and Remote Sensing*, 48(1):373–385, 2010. doi: https://doi.org/10.1109/TGRS.2009.2025498.

P.-E. Ng and K.-K. Ma. A switching median filter with boundary discriminative noise detection for extremely corrupted images. *IEEE Transactions on Image Processing*, 15(6):1506–1516, 2006. doi: https://doi.org/10.1109/TIP.2005.871129.

W. Ni and X. Gao. Despeckling of SAR image using generalized guided filter with Bayesian nonlocal means. *IEEE Transactions on Geoscience and Remote Sensing*, 54(1):567–579, 2016. ISSN 0196-2892. doi: https://doi.org/10.1109/TGRS.2015.2462120.

L. Pardo, D. Morales, M. Salicrú, and M. L. Menéndez. Generalized divergence measures: information matrices, amount of information, asymptotic distribution, and its applications to test statistical hypotheses. *Information Sciences*, 84(8):181–198, 1995. doi: https://doi.org/doi:10.1016/0020-0255(95)00017-J.

L. Pardo, D. Morales, M. Salicrú, and M. L. Menéndez. Large sample behavior of entropy measures when parameters are estimated. *Communications in Statistics – Theory and Methods*, 26(2):483–501, 1997. doi: https://doi.org/10.1080/03610929708831929.

K. Pearson. Contributions to the mathematical theory of evolution II: skew variation in homogeneous material. *Philosophical Transactions of the Royal Society A: Mathematical,*

Physical and Engineering Sciences, 186:343–414, 1895. doi: https://doi.org/10.1098/rsta.1895.0010.

K. Pearson. Das Fehlergesetz und Seine Verallgemeiner-ungen durch Fechner und Pearson: a rejoinder. *Biometrika*, 4(1–2):169–212, 1905. ISSN 0006-3444. doi: https://doi.org/10.1093/biomet/4.1-2.169.

P. A. A. Penna and N. D. A. Mascarenhas. SAR speckle nonlocal filtering with statistical modeling of Haar wavelet coefficients and stochastic distances. *IEEE Transactions on Geoscience and Remote Sensing*, 57(9):7194–7208, 2019. doi: https://doi.org/10.1109/TGRS.2019.2912153.

R Core Team. *R: A Language and Environment for Statistical Computing*. R Foundation for Statistical Computing, Vienna, Austria, 2020. https://www.R-project.org/.

M. Salicrú, D. Morales, M. L. Menéndez, and L. Pardo. On the applications of divergence type measures in testing statistical hypotheses. *Journal of Multivariate Analysis*, 51(2):372–391, 1994.

C. A. N. Santos, D. L. N. Martins, and N. D. A. Mascarenhas. Ultrasound image despeckling using stochastic distance-based BM3D. *IEEE Transactions on Image Processing*, 26(6):2632–2643, 2017. doi: https://doi.org/10.1109/tip.2017.2685339.

M. Schmitt, L. H. Hughes, and X. X. Zhu. The SEN1-2 dataset for deep learning in SAR-optical data fusion. *ISPRS Annals of the Photogrammetry, Remote Sensing and Spatial Information Sciences*, IV(1):141–146, 2018. doi: https://doi.org/https://doi.org/10.5194/isprs-annals-IV-1-141-2018.

M. Silva, F. Cribari-Neto, and A. C. Frery. Improved likelihood inference for the roughness parameter of the GA0 distribution. *Environmetrics*, 19(4):347–368, 2008. doi: https://doi.org/10.1002/env.881. http://www3.interscience.wiley.com/cgi-bin/abstract/114801264/ABSTRACT.

M. Soumekh. *Synthetic Aperture Radar Signal Processing with Matlab Algorithms*. John Wiley & Sons, Inc., New York, 1 edition, 1999.

Y. Sun, L. Lei, D. Guan, X. Li, and G. Kuang. SAR image speckle reduction based on nonconvex hybrid total variation model. *IEEE Transactions on Geoscience and Remote Sensing*, 59(2):1231–1249, 2021. doi: https://doi.org/10.1109/TGRS.2020.3002561.

The Turing Way Community, B. Arnold, L. Bowler, S. Gibson, P. Herterich, R. Higman, A. Krystalli, A. Morley, M. O'Reilly, and K. Whitaker. *The Turing Way: A Handbook for Reproducible Data Science*. Zenodo, March 2019. doi: https://doi.org/10.5281/zenodo.3233986.

C. Tomasi and R. Manduchi. Bilateral filtering for gray and color images. In *Sixth International Conference on Computer Vision*, pages 839–846, 1998. doi: https://doi.org/10.1109/ICCV.1998.710815.

L. Torres, S. J. S. Sant'Anna, C. C. Freitas, and A. C. Frery. Speckle reduction in polarimetric SAR imagery with stochastic distances and nonlocal means. *Pattern Recognition*, 47:141–157, 2014. doi: https://doi.org/10.1016/j.patcog.2013.04.001.

E. R. Tufte. *The Visual Display of Quantitative Information*. Graphics Press, 2 edition, 2001.

S. Uhlmann and S. Kiranyaz. Integrating color features in polarimetric SAR image classification. *IEEE Transactions on Geoscience and Remote Sensing*, 52(4):2197–2216, 2014.

K. L. P. Vasconcellos, A. C. Frery, and L. B. Silva. Improving estimation in speckled imagery. *Computational Statistics*, 20(3):503–519, 2005. doi: https://doi.org/10.1007/BF02741311.

S. Vitale, D. Cozzolino, G. Scarpa, L. Verdoliva, and G. Poggi. Guided patchwise nonlocal SAR despeckling. *IEEE Transactions on Geoscience and Remote Sensing*, 57(9):6484–6498, 2019. ISSN 1558-0644. doi: https://doi.org/10.1109/TGRS.2019.2906412.

H. Wainer. Robust statistics: a survey and some prescriptions. *Journal of Educational Statistics*, 1(4):285–312, 1976.

H. Wickham. *Advanced R*. Taylor & Francis Inc., 2 edition, 2019. ISBN 0815384572. https://adv-r.hadley.nz/index.html.

J. L. Wilson and C. M. Botham. Three questions to address rigour in your proposal. *Nature*, 596:609–610, 2021. doi: https://doi.org/https://doi.org/10.1038/d41586-021-02286-z.

J. Wu, F. Liu, H. Hao, L. Li, L. Jiao, and X. Zhang. A nonlocal means for speckle reduction of SAR image with multiscale-fusion-based steerable kernel function. *IEEE Geoscience and Remote Sensing Letters*, 13(11):1646–1650, 2016. ISSN 1545-598X. doi: https://doi.org/10.1109/LGRS.2016.2600558.

D.-X. Yue, F. Xu, A. C. Frery, and Y.-Q. Jin. A generalized Gaussian coherent scatterer model for correlated SAR texture. *IEEE Transactions on Geoscience and Remote Sensing*, 58(4):2947–2964, 2020. doi: https://doi.org/10.1109/TGRS.2019.2958125.

D.-X. Yue, F. Xu, A. C. Frery, and Y.-Q. Jin. SAR image statistical modeling Part II: spatial correlation models and simulation. *IEEE Geoscience and Remote Sensing Magazine*, 9(1):115–138, 2021a. doi: https://doi.org/10.1109/MGRS.2020.3027609.

D.-X. Yue, F. Xu, A. C. Frery, and Y.-Q. Jin. SAR image statistical modeling Part I: single-pixel statistical models. *IEEE Geoscience and Remote Sensing Magazine*, 9(1):82–114, 2021b. doi: https://doi.org/10.1109/MGRS.2020.3004508.

H. Zhong, Y. Li, and L. Jiao. SAR image despeckling using Bayesian nonlocal means filter with sigma preselection. *IEEE Geoscience and Remote Sensing Letters*, 8(4):809–813, 2011. doi: https://doi.org/10.1109/LGRS.2011.2112331.

Index

SAR Image Analysis — A Computational Statistics Approach: With R Code, Data, and Applications, First Edition.
Alejandro C. Frery, Jie Wu, and Luis Gomez.

Companion website: www.wiley.com/go/frery/sarimageanalysis

Printed and bound by CPI Group (UK) Ltd, Croydon, CR0 4YY
16/04/2025
14658424-0002